SpringerBriefs in Education

Mansoor Niaz · Mayra Rivas

Students' Understanding of Research Methodology in the Context of Dynamics of Scientific Progress

 Springer

Mansoor Niaz
Epistemology of Science Group,
    Department of Chemistry
Universidad de Oriente
Cumaná, Sucre
Venezuela

Mayra Rivas
Unidad Educativa La Inmaculada
Cumaná, Sucre
Venezuela

ISSN 2211-1921   ISSN 2211-193X   (electronic)
SpringerBriefs in Education
ISBN 978-3-319-32039-7   ISBN 978-3-319-32040-3   (eBook)
DOI 10.1007/978-3-319-32040-3

Library of Congress Control Number: 2016935973

Printed on acid-free paper

This Springer imprint is published by Springer Nature
The registered company is Springer International Publishing AG Switzerland

Acknowledgments

A major source of inspiration for this work was the seminal work of Gerald Holton (Harvard University) with respect to the oil drop experiment and the Millikan–Ehrenhaft controversy. Furthermore his continuous support and advice was crucial for developing various parts of this research project.

Our students willingly cooperated and participated in the different activities related to this project. We would like to express our sincere thanks to the following members of our research group for providing criticism and advice: Luis A. Montes, Ysmandi Páez, Marniev Luiggi, Arelys Maza, Cecilia Marcano, and Johhana Ospina.

Mayra Rivas is grateful to her parents (Gladys & Ciro), husband (José), and children (Gabriel & Gregorio) for providing a loving environment that helped to keep working. Mansoor Niaz wishes to thank daughter (Sabuhi) and wife (Magda) for their love, patience, and understanding which were essential for completing this project.

A special word of thanks is due to Bernadette Ohmer, Publishing Editor at Springer (Dordrecht) and Marianna Pascale, Senior Editorial Assistant, for their support, coordination, and encouragement throughout the various stages of publication.

Contents

Abstract

Atomic structure forms an important part of high school chemistry courses in almost all parts of the world. Among other aspects, this topic deals with the atomic models of J.J. Thomson (based on cathode ray experiments), E. Rutherford (based on alpha particle experiments), N. Bohr (based on quantum theory), and the elementary electrical charge (based on Millikan's oil drop experiment). The objective of this study is to facilitate high school students' understanding of research methodology based on alternative interpretations of data, role of controversies, creativity, and the scientific method, in the context of the oil drop experiment. These aspects form an important part of the nature of science (NOS). This study is based on a reflective, explicit, and activity-based approach to teaching nature of science (NOS). In this respect, the oil drop experiment has been of particular interest to science educators for facilitating students' understanding of research methodology and the dynamics of scientific progress (i.e., NOS). This study is based on three groups of high school students (10th grade, 15–18-year olds) enrolled at a public school in Venezuela. One group (n = 33) was randomly designated as the Control and the other two as Experimental Group A (n = 33) and Experimental Group B (n = 38), respectively. All three groups were taught by the same instructor and participated in the following activities: *First week*: Instruction in the traditional expository manner on the following aspects of atomic structure: Thomson, Rutherford, and Bohr models of the atom and the Millikan oil drop experiment for determining the elementary electrical charge. At the end of the week, students were asked to draw a concept map based on how they perceived the development of scientific knowledge. *Second week*: All three groups responded to a three-item Pretest. Experimental Groups A and B were provided a Study Guide based on the scientific method and the Millikan–Ehrenhaft controversy with respect to the determination of the elementary electrical charge (see Appendix). Students were asked to read the Study Guide over the weekend and prepare for discussing it the following week. *Third week*: Experimental Group students (A and B) were subdivided into small groups and asked to present and discuss what they considered to be the principal ideas in the Study Guide. The instructor acted as a moderator and clarified issues. Study Guide generated considerable discussion. After this interactive session, students were asked to draw another concept map based on what they considered to be the most important aspects of scientific development. *Fourth week*: Both Control and Experimental Group (A and B) students responded to a five-item Posttest. During the next month, 17 students from the Experimental Groups (A and B) and 11 from the Control Group were selected randomly for a semi-structured interview. Results obtained show that the difference in the performance (conceptual responses) of the Control and Experimental Group (A and B) students on the three items of the Pretest is statistically not significant. However, on the five items of the Posttest Experimental Groups performed better than the Control Group and the difference in the performance on conceptual responses is statistically significant ($p < 0.01$). After the experimental treatment most students changed their perspective and drew concept maps in which they emphasized the creative, accumulative, controversial nature of science and the scientific method. Interviews with students provided a good opportunity to observe how students' thinking changed after the experimental treatment. Multiple data sources were an important feature of this study. It is concluded that a teaching strategy based on a reflective, explicit, and activity-based approach in the context of the oil drop experiment can facilitate high school students' understanding of how scientists elaborate theoretical frameworks, design experiments, report data that leads to controversies and finally with the collaboration of the scientific community a consensus is reached.

Keywords History and philosophy of science · Historical reconstruction · Nature of science · Research methodology · Multiple data sources · Dynamics of scientific progress · Concept maps · Atomic structure · Atomic models of Thomson, Rutherford, and Bohr · Determination of the elementary electrical charge · Oil drop experiment · Millikan–Ehrenhaft controversy.

Introduction

How do scientists practice science? Do scientists depend on previous work of a topic? How do scientists interpret/understand experimental data? Do scientists disagree with respect to the interpretation of the same or similar experimental data? Do controversies and creativity play an important role in scientific progress? Is the scientific method important for doing scientific research? These are important questions if we want our students to understand the scientific endeavor. A review of the literature, however, shows that both the textbooks and science curricula generally ignore such questions (Niaz 2014). This study attempts to provide answers to such questions based on a historical reconstruction of the events that led to the determination of the elementary electrical charge (Holton 1978). This topic along with the atomic models of J.J. Thomson, E. Rutherford, and N. Bohr forms an important of understanding atomic structure both at the high school and introductory university level courses (Niaz 1998). Based on atomic structure, the need to facilitate students' conceptual understanding of the *particulate nature of matter* has been recognized by the NRC (2012). The history of the structure of the atom since the late nineteenth and early twentieth century shows that the atomic models of Thomson, Rutherford, Bohr, and wave mechanical evolved in quick succession and had to contend with competing models based on rival research programs. The emergence of these models and the ensuing controversies between the protagonists provide an illustration of an important aspect of the nature of science, namely the tentative nature of scientific knowledge. Furthermore, providing students with an insight with respect to how and why atomic models change provides an understanding of the dynamics of scientific progress (Niaz 2009). According to Trevor Levere (2006), a historian of chemistry:

> ...many authors of science textbooks still write as if there were such a thing as *the* scientific method, and use labels like induction, empiricism, and falsification in simplistic ways that bear little relation to science as it is practiced (pp. 115–116, original italics and underline added).

Science as it is practiced can indeed be an important guideline for science textbooks and teaching science (cf. science in the making, Niaz 2012). In order to implement this approach it is essential to present science content (atomic structure in this study) within a history and philosophy of science (HPS) perspective.

The objective of this study is to facilitate high school students' understanding of scientific research methodology based on alternative interpretations of data, role of controversies, creativity, and the scientific method. These aspects form an important part of the nature of science (NOS).

References

Holton, G. (1978). Subelectrons, presuppositions, and the Millikan-Ehrenhaft dispute. *Historical Studies in the Physical Sciences, 9,* 16–224.

Levere, T. H. (2006). What history can teach us about science: Theory and experiment, data and evidence. *Interchange, 37,* 115–128.

National Research Council, NRC (2012). *Discipline-Based Education Research: Understanding and Improving Learning in Undergraduate Science and Engineering.* Washington, DC: The National Academies Press.

Niaz, M. (1998). From cathode rays to alpha particles to quantum of action: a rationalreconstruction of structure of the atom and its implications for chemistry textbooks. *Science Education, 82,* 527–552.

Niaz, M. (2009). *Critical appraisal of physical science as a human enterprise: Dynamics of scientific progress.* Dordrecht, The Netherlands: Springer.

Niaz, M. (2012). *From 'science in the making' to understanding the nature of science: An overview for science educators.* New York: Routledge.

Niaz, M. (2014). Science textbooks: The role of history and philosophy of science. In M. R. Matthews (Ed.), *International handbook of research in history, philosophy and science teaching* (Vol. II, pp. 1411–1441). Dordrecht, The Netherlands: Springer.

Theoretical Framework

Nature of Science

The objective of helping students to develop informed views of the nature of science (NOS) has been a central goal for science education for the last many years (Abd-El-Khalick and Lederman 2000; Blanco and Niaz 2014; Clough and Olson 2004; Hodson and Wong 2014; Lederman et al. 2014; McComas 2008, 2014; Niaz 2012; Osborne et al. 2003; Smith and Scharmann 2008). The essence of NOS deals with how scientists do science, namely the:

1. Scientific knowledge is empirical and relies heavily on experimental evidence;
2. Relationship between experiment, data, and theory;
3. Role played by scientists' prior beliefs and presuppositions while they are designing new experiments;
4. Interpretation of the same experimental data in different ways;
5. Rivalries and conflicts among scientists as they conduct experiments and understand data;
6. Continual critical appraisal of theories leads to the tentative nature of scientific knowledge;
7. Objectivity in science as a social process of competitive cross-validation through critical peer review (cf. Campbell 1988a, b; Daston and Galison 2007; Phillips and Burbules 2000).

It is important to note that given the complexity and multifaceted nature of the issues involved and a running controversy among philosophers of science themselves, implementation of NOS in the classroom has also been difficult. Despite the controversy, a certain degree of consensus has been achieved within the science education community with respect to the seven issues outlined above.

Again, there has been some discussion in science education research as to the design of teaching strategies for introducing NOS in the classroom. However, there seems to be some consensus that teaching NOS needs a reflective, explicit, and

© The Author(s) 2016
M. Niaz and M. Rivas, *Students' Understanding of Research Methodology in the Context of Dynamics of Scientific Progress*, SpringerBriefs in Education, DOI 10.1007/978-3-319-32040-3_1

activity-based approach (Abd-El-Khalick and Akerson 2004; Akerson et al. 2006; Chang 2011; Ford and Wargo 2007; Hötteche et al. 2012; Khishfe and Lederman 2006, 2007; Niaz 2012; Smith and Scharmann 2008; Windschitl 2004; Wong et al. 2008). This study is based on a reflective, explicit, and activity-based approach to introducing NOS in the classroom.

Indeed, most science curricula and textbooks reduce 'scientific practice' to a simple accumulation of experimental data. Let us consider two examples from the history of science to show that such reduction does not facilitate students' understanding of scientific practice and consequently that of NOS. Most textbooks in almost all parts of the world report Rutherford's (1911) alpha particle experiments, which led to the postulation of the nuclear model of the atom. However, most textbooks ignore that J.J. Thomson (Rutherford's teacher and colleague) at about the same time conducted very similar alpha particle experiments at the Cavendish Laboratory in Cambridge University. Although both Rutherford and Thomson found very similar experimental results, still their interpretations were entirely different which led to a bitter dispute between the two protagonists that lasted for many years (for details, see Niaz 2009; Wilson 1983). Rutherford postulated the hypothesis of single scattering, whereas Thomson postulated the hypothesis of compound scattering. This shows that a 'simple inspection of phenomena' did not help Thomson and Rutherford to resolve the controversy and thus understand the experimental data.

Another example is provided by experimental data that led to the determination of the elementary electrical charge by R. Millikan and F. Ehrenhaft in the period 1909–1925 (Holton 1978; Niaz 2005). Although both researchers had very similar experimental data, inspection of phenomena was far from simple, as Millikan postulated the existence of a universal electrical charge (the electron) and Ehrenhaft postulated the existence of fractional electrical charges (subelectrons). More details of the Millikan–Ehrenhaft controversy are provided in a later section (also see Appendix, Study Guide).

These two controversies illustrate the role played by controversies in scientific progress which has been recognized in the history and philosophy of science literature:

> What is not so obvious and deserves attention is a sort of paradoxical dissociation between science as actually practiced and science as perceived or depicted by both scientists and philosophers. While nobody would deny that science in the making has been replete with controversies, the same people often depict its essence or end product as free from disputes, as the uncontroversial rational human endeavor par excellence (Machamer et al. 2000, p. 3).

Science educators and textbooks generally emphasize the 'end product' and hence generally ignore the role played by controversies.

Research in science education shows that teaching NOS involves two aspects, namely domain-general and domain-specific. Table 1 provides some examples to understand the difference (for further elaboration see Niaz 2016).

Table 1 Relationship between domain-general and domain-specific aspects of NOS[a]

Domain general	Domain specific
Empirical	Determination of mass-to-charge ratio of cathode rays/oil drop experiment
Rival theories	Valence bond and molecular orbital models of chemical bonding/Copenhagen, Schrödinger, and de Broglie hypotheses of quantum mechanics
Alternative interpretations	Alpha particle experiments/oil drop experiment
Theory-laden	Determination of elementary electrical charge: Millikan's and Ehrenhaft's presuppositions
Tentative	Atomic models in the twentieth century/from Newtonian mechanics to Einstein's theory of relativity
Objectivity	Alpha particle experiments/oil drop experiment/bending of light in the 1919 eclipse experiments
Social and historical milieu	Oil drop experiment/Michelson–Morley experiment

[a]This is a selected list of NOS aspects, to provide an overview. More detailed information can be found in Lederman et al. (2002), McComas et al. (1998), Niaz (2009, 2012, 2016)

Historical Reconstruction of the Oil Drop Experiment

For understanding the oil drop experiment within a domain-general NOS framework, it is essential that students be provided with the social and historical milieu. In other words, students need to know the context in which it was conducted, that means the interactions between the protagonists (Millikan and Ehrenhaft), that led to controversies and their respective theoretical frameworks which facilitated an understanding of what the experiment was all about. This study attempts to facilitate students' understanding of the social and historical context in which the oil drop experiment was conducted. At this stage, it is important to recognize that both the domain-general and domain-specific aspects of NOS complement each other and are essential for facilitating students' conceptual understanding based on an integration of the two aspects. Information included in this section facilitated the elaboration of the Study Guide (see Appendix).

The oil drop experiment has been the subject of considerable research (Holton 1978, 1999, 2014; Niaz 2005, 2015, 2016). Examination of Millikan's two laboratory notebooks by Holton (1978) revealed that Millikan (1913) studied 140 oil drops and published data for only 58 drops. Interestingly, Millikan (1913) meticulously presented complete data on 58 drops and emphasized: 'It is to be remarked, too, that this is not a selected group of drops but represents all of the drops experimented upon during 60 consecutive days …' (p. 138, original italics). If Millikan had included all the 140 oil drops (many of which had experimental errors), he would have obtained a wide range of values for the elementary electrical charge (*e*, the electron), similar to Ehrenhaft. At this stage, it is interesting to consider Holton's (1978) comment on this: 'If Ehrenhaft had obtained such data, he would probably not have neglected the second observations and many others like it in these two notebooks that shared the same fate; he would very likely have used them all' (pp. 209–210). It is important to note that the controversy had an important underlying theoretical assumption, namely Millikan believed in a discrete elementary electrical charge (electron), whereas Eherenhaft was guided by the anti-atomist views of Mach and others that entailed fractional charges (subelectron). This in our opinion is the crux of the issue in the Millikan–Ehrenhaft controversy. From an educational point of view, it is important to note that Millikan (1913) did not follow the scientific method (it is not important that he did not refer to his handling of the data as such). Both Millikan and Ehrenhaft found data that provided evidence for a wide range of electrical charges. Millikan's notebooks revealed a different story, namely he did not follow what is generally considered to be the scientific method. With this background, it is plausible to suggest (as we did in the Study Guide, see Appendix) that Ehrenhaft followed some form of the scientific method, whereas Millikan did not. The fact that Ehrenhaft did not leave laboratory notebooks is not an issue as many scientists do not do so.

At this stage, it is important to consider how the science education community has referred to this aspect of the Millikan–Ehrenhaft controversy. Klassen (2009) who has worked with the oil drop experiment with modern apparatus concluded:

> The conflict between Millikan and Ehrenhaft presents a unique opportunity to highlight the complex nature of how science operates in a particular setting and provides students with a basis upon which to begin thinking about the nature of science. **Ehrenhaft's actions were guided by the traditional scientific method**, whereas Millikan's actions were guided by his presuppositions about electrons (Rodríguez and Niaz 2004). It is hoped that students who are exposed to this material will become more cognizant of how *they* do science and that they will begin to reflect on whether they are (or should be) guided by the traditional 'scientific method' or their presuppositions (p. 601, original italics, emphasis added).

It is important to note that Klassen (2009) explicitly refers to Ehrenhaft's work to be based on the traditional scientific method and thus endorses our interpretation.

Kolstø (2008) while recommending history of science for democratic citizenship has also referred to the Millikan–Ehrenhaft controversy and concluded:

> Both Millikan and Ehrenhaft struggled to produce accurate measurements, but with different guiding hypotheses. Based on his empiricist view, Ehrenhaft seems to have accepted all measurements obtained. Based on his belief in the existence of unobservable electrons, a realist view of scientific knowledge, Millikan was led to look for errors and uncertainties and adjust methods instead of judging his hypothesis as falsified (p. 984) … A superficial treatment of the history of science might easily result in 'fictionalized idealizations' … where currently accepted views are seen as the most high-grade, leaving little recognition and respect for the *creativity of scientists of the past* (p. 993, italics added).

Creativity of scientists of the past, in this context, refers to Millikan's creativity in handling of his data. Indeed, besides his experimental acumen, Millikan explored alternatives in the interpretation of his data. Paraskevopoulou and Koliopoulos (Paraskekevopoulou and Koliopoulos 2011) designed a study to understand nature of science through the Millikan–Ehrenhaft controversy and concluded: 'The role played by imagination and creativity is apparent in the fact that Millikan continuously improved his experimental methods when he could see "an individual electron riding on a drop of oil" and in identifying the possible sources of error that prevailed over the results of an experiment' (p. 947).

According to Silverman (1992): 'Ehrenhaft accepted all measurements in the belief that constituted *objective observation*' (p. 169, italics added). Objective observation in this context approximates to some form of the scientific method. At this stage, it is important to note that Ehrenhaft was a very competent experimental physicist and according to Holton (1978), '… there was never a direct laboratory disproof of Ehrenhaft's claims' (p. 220). In other words, just like Ehrenhaft, other scientists (including Millikan) also found a wide range of charges in the oil drop experiment. Precisely, according to Daston and Galison (2007), understanding of the data from the oil drop experiment required some form of 'scientific judgment' that led Millikan to exclude experimental data.

In their textbook, Olenick et al. (1985) reproduced the following quote from Millikan's laboratory notebook (dated 15 March, 1912; see Holton 1978 for Millikan's laboratory notebooks): 'One of the best ever [data] … almost exactly *right*. Beauty—publish' (original italics). After reproducing the quote, the textbook authors asked a very thought-provoking question: 'What's going on here? How can it be right if he's supposed to be measuring something he doesn't *know*? One might expect him to publish everything!' (p. 244, original italics). These are important issues related to nature of science, namely can a scientist know beforehand what he is going to find and what is even more difficult to understand is that how can a scientist know the right answer before doing the experiment. Interestingly, the authors themselves provided further insight and advice for students:

Now, you shouldn't conclude that Robert Millikan was a bad scientist … What we see instead is something about how real science [cutting-edge] is done in the real world. What Millikan was doing was not cheating. He was applying scientific judgment … But experiments must be done in that way. Without that kind of judgment, the journals would be full of mistakes, and we'd never get anywhere. So, then, what protects us from being misled by somebody whose 'judgment' leads to wrong results? Mainly, it's the fact that someone else with a different prejudice can make another measurement … Dispassionate, unbiased observation is supposed to be the hallmark of the scientific method. Don't believe everything you read. Science is a difficult and subtle business, and there is no method that assures success (Olenick et al. 1985, p. 244).

This highlights important issues in the interpretation of data from the oil drop experiment, namely observations are difficult to interpret and do need some form of creativity (and not necessarily the scientific method), based on scientific judgment.

In this section, we have provided evidence from a wide range of sources such as Klassen (2009), Kolstø (2008), Paraskevopoulou and Koliopoulos (2011), Silverman (1992), and Olenick et al. (1985), to show that Millikan did not interpret his data based on the scientific method and this required him to be more creative. Reference to these and similar issues in the Study Guide is based on details provided in this section.

From 'Science in the Making' to Contextual Teaching (Science Stories)

Clough and Olson (2004) have argued that the use of short stories in combination with a scientific dispute as a strategy for teaching aspects of NOS can improve students' understanding. According to Klassen (2006):

> School science lacks the vitality of investigation, discovery, and creative invention that often accompanies *science-in-the-making* … The humanizing and clarifying influence of history of science brings the science to life and enables the student to construct relationships that would have been impossible in the traditional decontextualized manner in which science has been taught (p. 48, italics added).

In a sense, this contextual approach outlines the reflective, explicit, and activity-based approach to teaching nature of science (NOS). In this respect, the oil drop experiment has been of particular interest to science educators for facilitating students' understanding of research methodology and the dynamics of scientific progress (i.e., NOS). For example, Klassen (2009) deigned a study for Canadian undergraduate physics students to show the difficulty in obtaining results in cutting-edge experiments (e.g., oil drop), if the traditional scientific method is followed rather than allowing presuppositions to guide the interpretation of data. One of the students provided the following insight after this contextual and novel laboratory experience:

> If we had used the data for every drop we observed, our results would have not agreed with the accepted value at all. I suppose that Millikan must have depended quite heavily on his preconception of the value of e [elementary electrical charge], assuming his apparatus was similar to ours. If Millikan did not have a testable basis for rejecting drops …, I cannot see how the experiment would give one confidence that charge quantization had been observed (Reproduced in Klassen 2009, p. 604).

In another study, Paraskevopoulou and Koliopoulos (2011) based on Greek high school students (16–17-year-olds) used the Millikan–Ehrenhaft dispute in the story format to facilitate their understanding of the following NOS aspects: (a) the role played by empirical data in scientific debate; (b) the distinction between observation and inference; (c) the role of the scientist's imagination and creativity in the elaboration of a theory; and (d) the natural sciences have a subjective content during the formation of a theory. Results obtained revealed that students' understanding of all four NOS aspects that were targeted improved significantly and the authors concluded: 'The students were given the opportunity to comprehend that the knowledge of the existence of an elementary electric charge is not an objective fact which cannot be doubted but is precisely a human invention that was subject to a public debate among specialists. It appears that the narration of the dispute helped the students to understand science's internal processes, the introduction of a new theory in particular and its relationship with the experiment' (p. 956).

References

Abd-El-Khalick, F., & Akerson, V. L. (2004). Learning about nature of science as conceptual change: Factors that mediate the development of preservice elementary teachers' views of science. *Science Education, 88*, 785–810.

Abd-El-Khalick, F., & Lederman, N. G. (2000). The influence of history of science courses on students' views of nature of science. *Journal of Research in Science Teaching, 37*, 1057–1095.

Akerson, V. L., Morrison, J. A., & Roth McDuffie, A. (2006). One course is not enough: Preservice elementary teachers' retention of improved views of nature of science. *Journal of Research in Science Teaching, 43*, 194–213.

Blanco, E., & Niaz, M. (2014). Venezuelan university students' understanding of the nature of science. *Journal of Science Education, 15*(2), 66–70.

Campbell, D. T. (1988a). The experimenting society. In E. S. Overman (*ed.*), *Methodology and epistemology for social science* (pp. 290–314). Chicago: University of Chicago Press (first published 1971).

Campbell, D. T. (1988b). Can we be scientific in applied social science? In E. S. Overman (Ed.), *Methodology and epistemology for social science* (pp. 315–333). Chicago: University of Chicago Press (first published 1984 in *Evaluation Studies Review Annual*), (pp. 315–333).

Chang, H. (2011). How historical experiments can improve scientific knowledge and science education: The cases of boiling water and electrochemistry. *Science and Education, 20*, 317–341.

Clough, M. P., & Olson, J. K. (2004). The nature of science: Always part of the science story. *The Science Teacher, 71*(9), 28–31.

Daston, L., & Galison, P. (2007). *Objectivity*. New York: Zone Books.

Ford, M., & Wargo, B. M. (2007). Routines, roles, and responsibilities for aligning scientific and classroom practices. *Science Education, 91*, 133–157.

Hodson, D., & Wong, S. L. (2014). From the horse's mouth: Why scientists' views are crucial to nature of science understanding. *International Journal of Science Education, 36*(16), 2639–2665.

Holton, G. (1978). Subelectrons, presuppositions, and the Millikan-Ehrenhaft dispute. *Historical Studies in the Physical Sciences, 9*, 161–224.

Holton, G. (1999). Personal communication to the first author, April 29.

Holton, G. (2014). Personal communication to the first author, August 3.

Höttecke, D., Henke, A., & Riess, F. (2012). Implementing history and philosophy in science teaching: Strategies, methods, results and experiences from the European HIPST Project. *Science and Education, 21*, 1233–1261.

Khishfe, R., & Lederman, N. G. (2006). Teaching nature of science within a controversial topic: Integrated versus non-integrated. *Journal of Research in Science Teaching, 43*, 395–418.

Khishfe, R., & Lederman, N. (2007). Relationship between instructional context and views of nature of science. *International Journal of Science Education, 29*, 939–961.

Klassen, S. (2006). A theoretical framework for contextual science teaching. *Interchange, 37*, 31–62.

Klassen, S. (2009). Identifying and addressing student difficulties with the Millikan oil drop experiment. *Science and Education, 18*, 593–607.

Kolstø, S. D. (2008). Science education for democratic citizenship through the use of history of science. *Science and Education, 17*, 977–997.

Lederman, N. G., Abd-El-Khalick, F., Bell, R. L., & Schwartz, R. (2002). Views of nature of science questionnaire: Toward valid and meaningful assessment of learners' conceptions of nature of science. *Journal of Research in Science Teaching, 39*, 497–521.

Lederman, N. G., Bartos, S. A., & Lederman, J. S. (2014). The development, use, and interpretation of nature of science assessments. In M. R. Matthews (Ed.), *International handbook of research in history, philosophy and science teaching* (Vol. II, pp. 971–997). Dordrecht: Springer.

Machamer, P., Pera, M., & Baltas, A. (2000). Scientific controversies: An introduction. In P. Machamer, M. Pera, & A. Baltas (Eds.), *Scientific controversies: Philosophical and historical perspectives* (pp. 3–17). New York: Oxford University Press.

McComas, W. F. (2008). Seeking historical examples to illustrate key aspects of the nature of science. *Science and Education, 17*(2), 249–263.

McComas, W. F. (2014). Nature of science in the science curriculum and in teacher education programs in the United States. In M. R. Matthews (Ed.), *International handbook of research in history, philosophy and science teaching* (Vol. III, pp. 1993–2023). Dordrecht: Springer.

McComas, W. F., Almazroa, H., & Clough, M. P. (1998). The role and character of the nature of science in science education. *Science and Education, 7*, 511–532.

Millikan, R. A. (1913). On the elementary electrical charge and the Avogadro constant. *Physical Review, 2*, 109–143.

Niaz, M. (2005). An appraisal of the controversial nature of the oil drop experiment: Is closure possible? *British Journal for the Philosophy of Science, 56*, 681–702.

Niaz, M. (2009). *Critical appraisal of physical science as a human enterprise: Dynamics of scientific progress*. Dordrecht, The Netherlands: Springer.

Niaz, M. (2012). *From 'science in the making' to understanding the nature of science: An overview for science educators*. New York: Routledge.

Niaz, M. (2015). Myth 19: That the Millikan oil-drop experiment was simple and straightforward. In R. L. Numbers & K. Kampourakis (Eds.), *Newton's Apple and Other Myths about Science* (pp. 157–163). Cambridge, MA: Harvard University Press.

Niaz, M. (2016). *Chemistry education and contributions from history and philosophy of science*. Dordrecht: Springer.

Olenick, R. P., Apostol, T. M., & Goodstein, D. L. (1985). *Beyond the mechanical universe. From electricity to modern physics*. New York: Cambridge University Press.

Osborne, J., Collins, S., Ratcliffe, M., Millar, R., & Duschl, R. (2003). What 'ideas-about-science' should be taught in school science? A Delphi study of the expert community. *Journal of Research in Science Teaching, 40*, 692–720.

Paraskekevopoulou, E., & Koliopoulos, D. (2011). Teaching the nature of science through the Millikan-Ehrenhaft dispute. *Science and Education, 20*(10), 943–960.

Phillips, D. C., & Burbules, N. C. (2000). *Postpositivism and educational research*. New York: Rowman & Littlefield.

Rodríguez, M. A., & Niaz, M. (2004). The oil drop experiment: An illustration of scientific research methodology and its implications for physics textbooks. *Instructional Science, 32*, 357–386.

Rutherford, E. (1911). The scattering of alpha and beta particles by matter and the structure of the atom. *Philosophical Magazine, 21*, 669–688.

Silverman, M. P. (1992). Raising questions: Philosophical significance of controversy in science. *Science and Education, 1*, 163–179.

Smith, M. U., & Scharmann, L. C. (2008). A multi-year program developing an explicit reflective pedagogy for teaching pre-service teachers the nature of science by ostention. *Science and Education, 17*, 219–248.

Wilson, D. (1983). *Rutherford: Simple genius*. Cambridge, MA: MIT Press.

Windschitl, M. (2004). Folk theories of "inquiry:" How preservice teachers reproduce the discourse and practices of an atheoretical scientific method. *Journal of Research in Science Teaching, 41*, 481–512.

Wong, S. L., Hodson, D., Kwan, J., & Wai Jung, B. H. (2008). Turning crisis into opportunity: Enhancing student-teachers' understanding of nature of science and scientific inquiry through a case study of the scientific research in severe acute respiratory syndrome. *International Journal of Science Education, 30*(11), 1417–1439.

Method

This study is based on three groups of high school students (10th grade, 15–18-year-olds) enrolled at a public school in Venezuela. One group (n = 33) was randomly designated as the control and the other two as Experimental Group A (n = 33) and Experimental Group B (n = 38), respectively. The number of students on some test items varies as all the students registered in the course did not attend class on that day. All three groups were taught by the same instructor (second author of this study). Following is a summary of the activities in which the three groups participated:

First week: All three groups received instruction in the traditional expository manner on the following aspects of the topic on atomic structure: Thomson, Rutherford, and Bohr models of the atom and the Millikan oil drop experiment for determining the elementary electrical charge. This presentation was based on the textbook by Caballero and Ramos (2001), which follows the traditional approach characterized as 'rhetoric of conclusions' (Schwab 1974); namely, students were not allowed to reason and understand the underlying arguments. Most of the time during this week was spent on emphasizing (based on the textbook approach) experimental details (Thomson, Rutherford, Millikan and in the case of Bohr experimental evidence related to hydrogen line spectra) and ignoring the rationale of why and how the scientist was doing his work. At the end of the week, students were asked to draw a concept map based on how they perceived the development of scientific knowledge. Students had received instruction on the elaboration of concept maps in a previous course based on the work of J. Novak (Ausubel et al. 1991; Novak and Gowin 1988; Novak 1990).

Second week: All three groups responded to a 3-item Pretest (presented later in this section). Experimental Groups A and B were provided a Study Guide based on the scientific method and the Millikan–Ehrenhaft controversy with respect to the determination of the elementary electrical charge (see Study Guide, Appendix). Students were asked to read the Study Guide over the weekend and prepare for discussing it the next week. During this week, Control Group students continued to discuss the atomic models of Thomson, Rutherford, and Bohr. On the other hand,

© The Author(s) 2016

M. Niaz and M. Rivas, *Students' Understanding of Research Methodology in the Context of Dynamics of Scientific Progress*, SpringerBriefs in Education, DOI 10.1007/978-3-319-32040-3_2

Experimental Group students were asked to read the Study Guide, followed by a question–answer session dealing with various aspects of the Millikan–Ehrenhaft controversy.

Third week: Experimental Group students (A and B) were subdivided into small groups and asked to present and discuss what they considered to be the principal ideas in the Study Guide. The instructor acted as a moderator and clarified issues. Discussion of the Study Guide generated considerable discussion and following are some of the salient features: (a) How could Millikan discard data and not report it in his published paper? (b) It is not clear how Ehrenhaft followed the scientific method and still lost support of the scientific community? and (c) Do all scientists work like Millikan and Ehrenhaft? After this interactive session, students were asked to draw another concept map based on what they considered to be the most important aspects of scientific development. During this period, the Control Group students were provided instruction with respect to the atomic structure based on the traditional methodology. For example, simple problems based on electronic transitions (Balmer formula) were solved.

Fourth week: Both Control and Experimental Group (A and B) students responded to a 5-item Posttest. Besides this, all groups received instruction with respect to a simple version of the wave mechanical model of the atom, uncertainty principle, quantum numbers, and electron configurations of chemical elements.

During the next month, 17 students from the Experimental Groups (A and B) and 11 from the Control Group were selected randomly for a semi-structured interview, which was conducted by the second author. All interviews were audiotaped and then transcribed. Each interview lasted about 30 min, and the instructor showed the students their responses to items from the Pretest and Posttest and requested clarification or any additional comments. Items in the Pretest and Posttest were elaborated by consulting the science education research literature and three investigators working in history and philosophy of science at our university.

Pretest

1. What in your opinion was Millikan's major contribution in the oil drop experiment? Did Millikan develop his experiment without receiving any contribution from other scientists?
2. After the scientists have developed a theory (for example atomic theory), does the theory ever change? Rutherford completely changed Thomson's model, and Bohr changed Rutherford's model. Do you think that these scientists made mistakes while doing the experiments?
3. How does scientific knowledge develop? Explain.

Posttest

1. Scientists have a unique method (scientific method) for carrying out their experiments; that is, there exists only one way of doing science. Can a diversity of methods exist for developing scientific knowledge? Explain.
2. Some astrophysicists believe that the universe is expanding, whereas others believe that it is in a static state with no expansion or reduction. How are these

different conclusions possible if all of these scientists have done the same experiments and have the same experimental data?

3. Scientists do experiments in order to collect evidence to find answers to the hypotheses they have proposed. What is the importance of experimental data for scientists?

4. In your opinion, during the development of the experiments, does controversy with other scientists and creativity can help in the development of science?

5. In your opinion, what are the most important aspects of scientific development?

Item 2 of the Pretest and Items 1 and 2 of the Posttest are adapted from Lederman et al. (2002) as part of their *VNOS-Form B* (p. 505). With respect to the scientific method, these authors stated: 'The myth of the scientific method is regularly manifested in the belief that there is a recipe like stepwise procedure that all scientists follow when they do science' (Lederman et al. 2002, p. 501). At this stage, it is important to note an important difference between items in the Pretest and Posttest of this study and those in *VNOS-Form B*. All items in this study are context specific or in other words have a domain-specific background. For example, Items 1 and 2 of the Pretest specifically refer to Millikan's oil drop experiment, Thomson, Rutherford, and Bohr's models of the atom—all these formed part of the students' chemistry course and were included in the textbook they followed (Caballero and Ramos 2001). This textbook formed part of a study that reported the evaluation of atomic structure in Venezuelan high school chemistry textbooks within a history and philosophy of science framework (Páez et al. 2004). Similarly, the textbook by Caballero and Ramos (2001) also formed part of a study that evaluated the oil drop experiment in Venezuelan high school chemistry textbooks (cf. López 2006). In the case of the Posttest, Item 2 refers to the work of the astrophysicists, whereas the other four items refer to the oil drop experiment.

Validation of Students' Responses on Items in Pretest and Posttest

In all items of the Pretest and Posttest, students were provided the opportunity to express their views, opinions, reasons, understandings, and of course epistemological beliefs. At no stage in this study, we have claimed that students' responses (both Control and Experimental Groups) were fully representative and exhaustive of their thinking at a particular point during the evaluation. Students were at liberty to express their views to the extent that they considered necessary and essential. Furthermore, it is important to note that the format of these items is very different from multiple choice evaluations. Responses on all items in the Pretest and the Posttest were classified as: conceptual, rhetorical, and no response. Most of the criteria for classification were the same as used by Niaz et al. (2002) and Niaz and

Luiggi (2014). In general, a conceptual response showed an understanding of the underlying issues, whereas a rhetorical response simply reiterated the information provided (quite similar to what Schwab 1974 has referred to as a 'rhetoric of conclusions'). In general, conceptual responses provided plausible reasons that supported a particular stance/understanding of the issues being explored and were based on reflections and not memorization. On the other hand, rhetorical responses were generally prescriptive with little attempt to present arguments/reasons for adopting a particular position. Furthermore, rhetorical responses at times reiterated memorized textbook presentations related to theories and models, which consider that if a theory/model is replaced it means that the previous formulation was erroneous. According to current history and philosophy of science, theories/models are not right or wrong but differ in their heuristic or explanatory power (cf. Lakatos 1970). Examples of both types of responses are provided in the next section. Responses of 4 students from the Control Group and 8 from the Experimental Groups (A & B) were classified by both authors. There was a coincidence of 76 % on the Pretest and 70 % on the Posttest. Disagreements were discussed in various meetings and both authors presented arguments and finally a consensus was achieved. Remaining responses were then classified by the second author, and in the case of further disagreements, both authors discussed and resolved the differences.

References

Ausubel, D., Novak, J., & Hanesian, H. (1991). *Psicología Educativa: Un punto de vista cognoscitivo*. México, D.F.: Trillas.

Caballero, A., & Ramos, F. (2001). *Química: Teoría, problemario, auto evaluación* (7th ed.). Caracas: Distribuidora Escolar.

Lakatos, I. (1970). Falsification and the methodology of scientific research programmes. In I. Lakatos & A. Musgrave (Eds.), *Criticism and the growth of knowledge* (pp. 91–195). Cambridge, UK: Cambridge University Press.

Lederman, N. G., Abd-El-Khalick, F., Bell, R. L., & Schwartz, R. (2002). Views of nature of science questionnaire: Toward valid and meaningful assessment of learners' conceptions of nature of science. *Journal of Research in Science Teaching, 39*, 497–521.

López, J. B. (2006). El enlace covalente y el experimento de Millikan, desde el punto de vista de la historia y filosofía de la ciencia, en libros de texto del primer año de ciencias del ciclo diversificado. Master of Science thesis (Chemistry education). Universidad de Oriente, Cumaná, Venezuela.

Niaz, M., & Luiggi, M. (2014). *Facilitating conceptual change in students' conceptual understanding of the periodic table*. Dordrecht, The Netherlands: Springer.

Niaz, M., Aguilera, D., Maza, A., & Liendo, G. (2002). Arguments, contradictions, resistances and conceptual change in students' understanding of atomic structure. *Science Education, 86*, 505–525.

Novak, J. D. (1990). Concept mapping: A useful tool for science education. *Journal of Research in Science Teaching, 27*(10), 937–949.

Novak, J., & Gowin, B. (1988). *Aprendiendo a aprender*. Madrid: Martínez Roca.

Páez, Y., Rodríguez, M. A., & Niaz, M. (2004). Los modelos atómicos desde la perspectiva de la historia y filosofía de la ciencia: Un análisis de la imagen reflejada por los textos de química de bachillerato. *Investigación y Postgrado, 19*(1), 51–77.

Schwab, J. J. (1974). The concept of the structure of a discipline. In E. W. Eisner & E. Vallance (Eds.), *Conflicting Conceptions of Curriculum* (pp. 162–175). Berkeley, CA: McCutchan Publishing Corp.

Results and Discussion

In this section, we report students' responses on seven items (Pretest and Posttest), concept maps drawn by the students before and after the experimental treatment, and the interviews with the students. Excerpts from Control and Experimental Group students are provided in order to facilitate students' understanding of the underlying issues. In general, the conceptual responses were more varied and indicate the extent to which students interacted with the experimental treatment or the context of a particular question. Rhetorical responses from Experimental Group students are not included as they were quite similar to those of the Control Group. This similarity between the rhetorical responses of both Control and Experimental Group students shows the difficulties involved in facilitating conceptual change.

Millikan and the Oil Drop Experiment (Students' Responses on Item 1 of Pretest)

The main objective of this item was to explore students' understanding of the oil drop experiment after having been exposed to a traditional presentation in the classroom as found in most high school chemistry textbooks (in this case, Caballero and Ramos 2001). A comparison of the performance of Control and Experimental Groups is presented in Table 1.

Results reported in Table 1 show that at the beginning of the course, the control and the two Experimental Groups have a very similar understanding of the oil drop experiment. Very few students provided conceptual responses, and a majority had a rhetorical response. Now, in order to understand better, let us see some examples of students' responses.

© The Author(s) 2016
M. Niaz and M. Rivas, *Students' Understanding of Research Methodology in the Context of Dynamics of Scientific Progress*, SpringerBriefs in Education, DOI 10.1007/978-3-319-32040-3_3

Table 1 Comparison of the performance of Control and Experimental Group (A and B) students on Item 1[a] (Pretest)

Response	Control ($n = 33$)	Experimental A ($n = 33$)	Experimental B ($n = 33$)
Conceptual	3 (9 %)	3 (9 %)	4 (12 %)
Rhetorical	26 (79 %)	19 (58 %)	23 (70 %)
No response	4 (12 %)	11 (33 %)	6 (18 %)

[a]Item 1: What in your opinion was Millikan's major contribution in the oil drop experiment? Did Millikan develop his experiment without receiving any contribution from other scientists?

Following is an example of a conceptual response by a Control Group student:

Millikan's major contribution was the determination of the charge of the electron to be 1.6×10^{-19} coulomb. He did consider the contribution of other scientists by using oil instead of water (Student #1, Control Group).

Following is an example of a conceptual response by an Experimental Group B student:

This experiment consisted in observing an oil drop between the two plates of a condenser which helped Millikan to determine the charge of the electron to be 1.6×10^{-19} coulomb. He considered the contribution of other scientists by changing water to oil drops (Student #5, Experimental Group B).

An example of a rhetorical response is provided by a student from the Control Group:

He [Millikan] determined that the charge of the electron was 1.60×10^{-19} coulomb (Student #23, Control Group).

Another example of a rhetorical response is provided by a student from the Experimental Group A:

He [Millikan] performed the oil drop experiment and determined the charge of the electron. He worked alone and did not look for support from others (Student #3, Experimental Group A).

Conceptual and the rhetorical responses by the two groups of students are very similar. Conceptual responses refer to the value of electron's charge as determined by Millikan and go beyond by pointing out that other scientists contributed to his understanding. On the other hand, rhetorical responses show no interest with respect to how Millikan determined the charge of the electron. Of course, as the Study Guide (see Appendix) shows the story behind the determination of the elementary electrical charge is complex and rich with details that can be of considerable interest to students. However, for these students, their conceptual understanding is based on the textbook they used (before the experimental treatment). Interestingly, the previous research has shown that even introductory university-level general chemistry textbooks ignore most of the details of this historical episode (for general chemistry textbooks published in USA see Niaz 2000; for textbooks published in Turkey see Niaz and Coştu 2013; and for chemistry high school textbooks published in Venezuela see López 2006).

Tentative Nature of Atomic Theories (Students' Responses on Item 2 of Pretest)

The main objective of this item was to evaluate students' understanding of the tentative nature of science (NOS) based on their understanding of atomic models/theories. Performance of the Control and Experimental Group students is presented in Table 2.

Results reported in Table 2 show that the control and the two Experimental Groups have a very similar understanding of the tentative nature of atomic theories. Very few students provided conceptual responses, and a majority had a rhetorical response. As compared to Item 1 (Pretest), the number of students who did not respond increased considerably. Now, in order to understand better, let us see some examples of students' responses.

Following is an example of a conceptual response by a Control Group student:

> As we can see in the case of atomic theory, one scientist completely changed the model of the previous. However, it does not mean that the previous scientist was wrong, but rather it shows that the following model is better defined (Student #4, Control Group).

Following is an example of a conceptual response by an Experimental Group A student:

> After the atomic theories have been developed these can change with the passage of time as other scientists do experiments to verify the theories that are still being used. We cannot say that these scientists made mistakes as Thomson, Rutherford and Bohr worked in different time periods (Student #1, Experimental Group A).

Now, let us compare these conceptual responses with rhetorical ones. Following is an example of a rhetorical response by a Control Group student:

> Theories cannot change as these have already been established (Student #18, Control Group).

Following is an example of a rhetorical response by an Experimental Group B student:

> Well, in my opinion scientists do make mistakes. Thomson made mistakes in his experiments and this was corrected by Rutherford. Similarly, Rutherford made mistakes and this was corrected by Bohr (Student #23, Experimental Group B).

Table 2 Comparison of the performance of Control and Experimental Group (A and B) students on Item 2[a] (Pretest)

Response	Control ($n = 33$)	Experimental A ($n = 33$)	Experimental B ($n = 33$)
Conceptual	2 (6 %)	3 (9 %)	6 (18 %)
Rhetorical	15 (45 %)	22 (67 %)	17 (52 %)
No response	16 (48 %)	8 (24 %)	10 (30 %)

[a]Item 2: After the scientists have developed a theory (e.g., atomic theory), does the theory ever change? Rutherford completely changed Thomson's model and Bohr changed Rutherford's model. Do you think that these scientists made mistakes while doing the experiments?

There is some ambiguity in these two rhetorical responses. Student #18 stated that the theories cannot change, but did not clarify whether these scientists made mistakes. Student #23 stated that these scientists did make mistakes but does not clarify whether scientific theories are tentative. Some other students also showed similar ambiguity. Comparing the conceptual and the rhetorical responses on Item 2 reveals that in order to respond correctly, the student must understand that scientists generally do not make mistakes while doing their experiments. After recognizing this facet of the scientific endeavor, the student has to go beyond and understand that atomic models changed as the subsequent models provided greater insight (based on new experiments) and consequently increased the explanatory power of the model. In other words, although scientists generally do not make mistakes while doing the experiments, their models are still open to criticism and change. Of course, this does not mean that the experimental work of the scientists and the subsequent model building is infallible. These results show the difficulties involved in understanding the tentative NOS.

Development of Scientific Knowledge (Students' Responses on Item 3 of Pretest)

The main objective of this item was to evaluate the degree to which students understand how scientific knowledge develops, based on their present and previous courses related to science subjects. Performance of the Control and Experimental Group students is presented in Table 3.

Results reported in Table 3 show once again that the control and the two Experimental Groups have a very similar understanding of the development of scientific knowledge. Very few students provided conceptual responses, and a majority had a rhetorical response.

Following is an example of a conceptual response by a Control Group student:

> Science studies the reality based on human beings, animals and the environment through experiments, hypotheses, results, and conclusions. Subsequently, this leads to the formulation of theories that help in the evolution of technology that contributes to our personal and social progress (Student #26, Control Group).

Table 3 Comparison of the performance of Control and Experimental Group (A and B) students on Item 3[a] (Pretest)

Response	Control ($n = 33$)	Experimental A ($n = 33$)	Experimental B ($n = 33$)
Conceptual	1 (3 %)	1 (3 %)	3 (9 %)
Rhetorical	19 (58 %)	24 (73 %)	19 (58 %)
No response	13 (39 %)	8 (24 %)	11 (33 %)

[a]Item 3: How does scientific knowledge develop? Please explain

Following is an example of a conceptual response by an Experimental Group B student:

Science advances by the work of the scientists who report their experiments and draw conclusions. While producing scientific knowledge, scientists also make mistakes that lead them to change their postulates (Student #15, Experimental Group B).

Now, let us compare these conceptual responses with rhetorical responses. Following is an example of a rhetorical response by a Control Group student:

Scientific development takes place in laboratories based on chemical processes that are indispensable for producing electric processes and medicines (Student #7, Control Group).

Following is an example of a rhetorical response by an Experimental Group A student:

From my point of view, in recent times scientific development has evolved considerably. Every day new medicines are discovered, based on the experiments and the untiring work of the scientists. This provides the whole world an opportunity to lead a better life free of diseases (Student #19, Experimental Group A).

It is interesting to note that on the conceptual responses, both the control and the Experimental Group students emphasize the importance of experiments, hypotheses, theories, and conclusions. Furthermore, while producing scientific knowledge scientists can make mistakes which lead them to change their postulates. In contrast, the rhetorical responses emphasize the role of experiments and laboratories and thus ignore the formulation of scientific theories. Both groups of students established a sort of direct relationship between the experiments and the technological advances, such as medicines and other goods.

At this stage, it is important to note that given the traditional science curriculum and the textbooks, it was not expected that the Control and Experimental Groups would differ in their performance on the three items of the Pretest. Also, it was expected that none or very few students would have conceptual responses. Actually, on the three items of the Pretest, there were 13 conceptual responses (Control Group = 6, Experimental Group = 7). It is plausible to suggest that the format of the items in the Pretest helped students to understand the underlying issues better which facilitated the conceptual responses. Consequently, results reported in Tables 1, 2, and 3 on the three items of the Pretest are quite concordant with the expectations of this research project. Now that we have a general idea of the scientific and NOS background of these students, it would be interesting to see how the Experimental Group's performance (Posttest) changes after the experimental treatment. Students responded to the five items of the Posttest after they had prepared concept maps (first week), gone through experimental treatment based on the Study Guide, Appendix (second week), and then prepared concept maps once again (third week).

Scientific Method (Students' Responses on Item 1 of Posttest)

The objective of this item was to explore students' views with respect to the scientific method after they had prepared concept maps, gone through the experimental treatment based on the Study Guide (see Appendix), and prepared concept maps once again.

Results reported in Table 4 show that none of the Control Group students responded conceptually. A majority (more than 80 % for both groups A and B) of the Experimental Group students responded conceptually. Most (97 %) of the Control Group students had a rhetorical response. Following are four examples of conceptual responses provided by Experimental Group students:

> Some scientists use the scientific method and there are others who do not use it, that is they use their experience, creativity and knowledge. There exist many methods for the development of scientific knowledge (Student #10, Experimental Group A).
>
> There is no unique scientific method, as scientists use their creativity and experience while doing the experiments. For example, Millikan used his knowledge and creativity to do the experiment (Student #15, Experimental Group A).
>
> All the scientists do not use the scientific method in their research. The method used depends on the nature of the research, theoretical framework of the scientist and his creativity (Student #27, Experimental Group B).
>
> Scientists do not have a unique scientific method for doing their experiments. There exist other methods, for example the one used by Millikan (Student #4, Experimental Group B).

It is important to note that these responses were considered as conceptual as they integrated various aspects of the information provided in the Study Guide (also classroom discussions) in order to support a particular position, namely the use of diverse methods by scientists. At this stage, it is interesting to compare these conceptual responses with rhetorical responses provided by the Control Group students:

> Various scientific methods do not exist. Scientists use just one method (Student #2, Control Group).
>
> There exist ways of doing science and for that we have to follow certain steps, such as observation, postulation of hypotheses, experimentation, analyses of the results to see if the experiment worked (Student #21, Control Group).

Table 4 Comparison of the performance of Control and Experimental Group (A and B) students on Item 1[a] of Posttest

Response	Control ($n = 32$)	Experimental A ($n = 33$)	Experimental B ($n = 38$)
Conceptual	– (–)	29 (88 %)	33 (87 %)
Rhetorical	31 (97 %)	4 (12 %)	5 (13 %)
No response	1 (3 %)	– (–)	– (–)

[a]Item 1: Scientists have a unique method (scientific method) for carrying out their experiments; that is, there exists only one way of doing science. Can a diversity of methods exist for developing scientific knowledge? Explain

> Yes, there exist diverse scientific methods, like the methods used by Thomson, Rutherford, Bohr and Millikan. Each one of them used the scientific method for the development of science (Student #17, Control Group).

Student #2 endorses the use of only one method. Student #21 follows the stepwise procedure outlined in most textbooks. Student #17, on the one hand, recognizes that a diversity of methods can exist and then reiterates that some of the scientists (Thomson, Rutherford & Bohr) used the scientific method. At this stage, it is important to note that rhetorical responses are not necessarily incorrect but rather obscure what the scientist actually did. For example, conceptual responses also acknowledge that scientists use some form of method, which is accompanied by the previous experience and creativity of the scientist.

Astrophysicists and the Expanding/Static Universe (Students' Responses on Item 2 of Posttest)

The objective of this item was to provide students an opportunity to respond to a question that was not discussed in class or in the textbook. For the Experimental Group students, this also meant that they could transfer knowledge from one context to another. For example, Experimental Group students discussed the Millikan–Ehrenhaft controversy (see Study Guide, Appendix) in class during the third week. This controversy showed that both Millikan and Eherenhaft had very similar experimental data and still their interpretations were entirely different. In a sense, the astrophysicists faced the same problem, namely having the same data some have been led to interpret data that leads to an expanding and others to a static universe (Table 5).

Most of the Experimental Group (A and B) students provided conceptual responses which is quite understandable. However, it was not expected that Control Group students would respond conceptually and still 16 % of these students did so. This is an interesting finding and will be discussed in this section. First let us see some examples of the conceptual responses provided by the Experimental Group students:

Table 5 Comparison of the performance of Control and Experimental Group (A and B) students on Item 2[a] (Posttest)

Response	Control ($n = 32$)	Experimental A ($n = 33$)	Experimental B ($n = 38$)
Conceptual	5 (16 %)	28 (85 %)	32 (84 %)
Rhetorical	26 (81 %)	5 (15 %)	6 (16 %)
No response	1 (3 %)	– (–)	– (–)

[a]Item 2: Some astrophysicists believe that the universe is expanding, whereas others believe that it is in a static state with no expansion or reduction. How are these different conclusions possible if all of these scientists have done the same experiments and have the same experimental data?

Astrophysicists interpret data that is also available to other scientists. However, both groups follow their own theories and thus do not arrive at the same conclusion. This does not mean that some of them might have made mistakes but rather indicates that they have different theoretical frameworks (Student #28, Experimental Group B).

It is not that the scientists [astrophysicists] made mistakes while doing the experiments. What leads them to formulate different interpretations is that they are guided by their own theories (Student #26, Experimental Group A).

Astrophysicists have the same data but they are guided by their own scientific knowledge and experience. This means that they critique the work of others and also share their knowledge with others, but still maintain their theoretical perspectives (Student #11, Experimental Group A).

They [astrophysicists] had the same data but each one of them elaborated his research following his own theoretical framework based on different methods. Similarly, Millikan did not use the scientific method, while Ehrenhaft strictly followed it and the two arrived at different conclusions (Student #7, Group B).

Now let us see the conceptual responses of the Control Group students:

The scientists [astrophysicists] can perform the same experiments and have the same data. However, they do not have the same form of approaching the problem nor the same knowledge. While doing research some can go deep and observe new things whereas others cannot do so (Student #3, Control Group).

These conclusions are difficult to understand. However, if we try to analyze a little we find that having the same data, what varies is that some follow the scientific method, while others use other methods, their creativity and knowledge (Student #13, Control Group).

Comparing the conceptual responses of the Control and Experimental Groups (A and B), it can be observed that (a) Experimental Group students use concepts such as 'theoretical frameworks,' 'theories,' and 'theoretical perspectives'; (b) Control Group students instead use concepts such as 'approaching the problem'; (c) Control Group students also refer to the 'scientific method' and 'creativity' that were discussed in the Study Guide (Experimental Groups), and it is possible that there could have been some out-of-class interaction between the Control and Experimental Groups. It is plausible to suggest that depending on the type of question (in this case based on astrophysicists), students' responses can manifest a 'testing effect' (cf. Lederman and O'Malley 1990, p. 236); that is, interaction with the question itself helps students to clarify their beliefs about the underlying issues and NOS. Furthermore, students' responses can also manifest a 'context effect' (Alexander 1992); that is, understanding improves if a question is embedded in a 'context' that is meaningful to the students. Item 2 of the Posttest based on astrophysicists and their varying interpretations of data are definitely thought-provoking and provide a 'context' that can facilitate greater understanding. In other words if a question elicits some reflection that can be helpful in arousing students' interest and creativity (cf. Niaz et al. 1991).

At this stage, it would be interesting to compare the conceptual responses (presented above) with the rhetorical responses. Following are some examples of rhetorical responses provided by Control Group students:

If they [astrophysicists] believe that the universe is expanding then all these scientists are wrong (Student #24, Control Group).

> I am in agreement with the scientists who believe that the universe is expanding and those who disagree have their own theories (Student #6, Control Group).
>
> Well, for me both groups of astrophysicists are wrong. Granted, that nobody is perfect, these scientists contrast their prior knowledge of the universe with their experimental results, and as it happens their experiments did not come out as they had expected them (Student #12, Control Group).

Most students who responded rhetorically simply reiterate one or the other thesis without providing arguments or a rationale (hence the rhetoric). A major difficulty with such responses is the lack of an understanding with respect to the controversial nature of all experimental data. In all scientific endeavors, the experimental data do not dictate the theory, but rather provide one of several different forms (based on a theoretical framework) of understanding the data. On the contrary, most science curricula and textbooks inculcate just the opposite, namely experimental data unambiguously lead the scientist to the correct theory.

Relationship Between Experimental Data and Scientific Theories (Students' Responses on Item 3 of Posttest)

As compared to Item 2 of the Posttest, this item deals with a domain-general aspect of the NOS, namely the interaction between experimental data and scientific theories. The main objective of this item was to explore actual historical episodes that can facilitate an understanding of the relationship between experimental data and scientific theories.

Table 6 shows that a majority of the Experimental Group students (A and B) provided conceptual responses, and none were classified as 'no response.' On the other hand, a majority of the Control Group students had rhetorical responses. Following are some examples of conceptual responses from Experimental Group students:

> For scientists, their theoretical frameworks are more important than experimental data. This is important as *data often lead to contradictions and conflicts* and in such situations it is the hypotheses which help scientists to design new experiments (Student #32, Experimental Group B, italics added).
>
> There is a great difference between the theoretical framework and the experimental data. It is the former that is more important for a scientist as it helps to analyze the data. Data are

Table 6 Comparison of the performance of Control and Experimental Group (A and B) students on Item 3[a] (Posttest)

Response	Control ($n = 32$)	Experimental A ($n = 33$)	Experimental B ($n = 38$)
Conceptual	1 (3 %)	26 (79 %)	34 (89 %)
Rhetorical	30 (94 %)	7 (21 %)	4 (11 %)
No response	1 (3 %)	– (–)	– (–)

[a]Item 3: Scientists do experiments in order to collect evidence to find answers to the hypotheses they have proposed. What is the importance of experimental data for the scientists?

important for providing *evidence for a hypothesis*, but the theoretical framework is even more important (Student #6, Experimental Group A, italics added).

Every scientist does experiments in order to find answers to the hypotheses and this constitutes evidence. Although, data are important the theoretical frameworks are essential. One example is provided by the experimental work of Millikan and Ehrenhaft, who had the same experimental data but different theoretical frameworks. This shows that every scientist uses his *creativity and knowledge to analyze the data* (Student #34, Experimental Group B, italics added).

The most important thing for a scientist is his theory — this helps them to *foresee where the work is heading*, and it makes easier to analyze the data (Student #27, Experimental Group A, italics added).

Experimental data are very important for the scientists as this enables them to follow the experiment. However, the theoretical framework of the scientists is even more important. For example, Millikan believed that an elementary electrical charge existed, whereas *Ehrenhaft believed in just the opposite* (Student #17, Experimental Group B, italics added).

These conceptual responses by the Experimental Group students show that they do understand fairly well the importance and the difference between experimental data versus theoretical frameworks. Student #32 rightly points out how *data can lead to contradictions and conflicts*, and it is precisely in such situations that the theoretical framework of the scientist plays an important role for designing new experiments. Student #6 refers to how the data provide *evidence for a hypothesis*. Student #34 refers to the Millikan–Ehrenhaft controversy (discussed in class), and how having the same experimental data, it is the *creativity and knowledge* of the scientist that make the crucial difference. Student #27 adds an important element in order to understand the difference between data and theoretical framework as the latter helps the scientist to *foresee where the work* [data] *is heading*. Student #17 highlights the fact that the theoretical frameworks of Millikan and Ehrenhaft were entirely different. This clearly shows that the Experimental Group students are not simply reproducing what was discussed in class (based on the Study Guide), but rather adding new elements in order to understand the difference between experimental data and theoretical frameworks. Some of these new elements are the following: Data can lead to contradictions and conflicts; data constitute evidence for hypotheses, creativity, and knowledge of the scientist in understanding data; a theoretical framework can help a scientist to foresee where the work is heading; scientists can have the same experimental data; and still their theoretical frameworks can be entirely different. These are important contributions toward understanding the difference between experimental data and theoretical frameworks that are generally not discussed in classrooms or science textbooks.

Following are some examples of rhetorical responses provided by the Control Group students:

Yes, scientists do experiments as these provide more knowledge with respect to the subject under study. This helps in going beyond what is already known (Student #27, Control Group).

For the scientists the data that they use are of utmost importance as this helps to find answers to the hypotheses (Student #5, Control Group).

All the scientists while doing the experiments base their work on: observations, proposing hypotheses, experimentation, results, and analyses. This helps them in collecting sufficient evidence from the experiment. In other words, through this *chain of steps* scientists obtain more information for finding answers to their hypotheses (Student #10, Control Group, italics added).

These rhetorical responses reproduce quite clearly what is discussed in most classrooms and science textbooks, namely experimentation as the major source of scientific development. Holton (1969) has referred to this aspect as the 'experimenticism fallacy.' Student #10 even refers to the *chain of steps* that correspond quite closely to the traditional *scientific method* found in most textbooks. The difference between the conceptual and rhetorical responses on this item clearly shows the role and importance of explicit and reflective approaches in understanding NOS (Akerson et al. 2000: also see Appendix, Study Guide).

Relationship Between Controversy, Creativity, and Progress in Science (Students' Responses on Item 4 of Posttest)

The main objective of this item was to highlight the role played by controversies during the development of an experiment and how in such situations the creativity of the scientists can help to facilitate understanding. Philosophers of science have recognized the role played by controversies in scientific progress in explicit terms:

> Many major steps in science, probably all dramatic changes, and most of the fundamental achievements of what we now take as the advancement or progress of scientific knowledge have been controversial and have involved some dispute or another. Scientific controversies are found throughout the history of science. This is so well known that it is trivial (Machamer et al. 2000, p. 3).

Perl (2007), Nobel Laureate in physics, has recognized the role of creativity in scientific progress in the following terms: 'An improvement of computer architecture, a discovery of a new medicine, a new understanding of the behavior of black holes, an improvement in gasoline engine efficiency—all are creative feats that are nonetheless limited by the laws of nature' (p. 2). This clearly shows the importance of creativity in a wide range of activities related to the scientific endeavor. In a similar vein, Lakatos (1970) has recognized that 'The direction of science is determined by human creative imagination and not by the universe of facts which surround us' (p. 187). Interestingly, the role of creativity in teaching NOS has also been recognized in science education research (e.g., Osborne et al. 2003). Similarly, the importance of the role played by creativity has been reported in various historical episodes by McComas (2008).

Table 7 shows that a majority of the Experimental Group students responded conceptually, whereas a majority of the Control Group students had rhetorical responses.

Table 7 Comparison of the performance of Control and Experimental Group (A and B) students on Item 4[a] (Posttest)

Response	Control (n = 32)	Experimental A (n = 33)	Experimental B (n = 38)
Conceptual	1 (3 %)	28 (85 %)	32 (84 %)
Rhetorical	28 (90 %)	3 (9 %)	6 (16 %)
No response	3 (9 %)	2 (6 %)	– (–)

[a]Item 4: In your opinion, during the development of an experiment, controversy among the scientists and their creativity can help in the progress of science?

Following are six examples of conceptual responses provided by Experimental Group students:

> Creativity of the scientists is very important in order to obtain results. Millikan and Ehrenhaft had a controversy as the former was *more creative and did not use the scientific method*. On the other hand, Ehrenhaft used the scientific method. Furthermore, both scientists followed their theoretical frameworks (Student #28, Experimental Group A, italics added).
>
> For me, creativity of the scientists can help in the development of science. For example, in Millikan's experiment as it was very difficult to keep the water drop under observation for more than a minute, he substituted water with oil. This helped him to build a new apparatus for the oil drop experiment. In this case Millikan used his *creativity to invent the experiment* (Student #3, Experimental Group B, italics added).
>
> The controversy between Millikan and Ehrenhaft provoked the *curiosity of the scientific community* and this contributed toward the development of science (Student #7, Experimental Group B, italics added).
>
> Creativity of the scientist is of great importance for scientific development, as it permits the researcher to have better ideas. Millikan used his *creativity to discover the oil drop experiment* that helped him to analyze the data adequately (Student #18, Experimental Group A, italics added).
>
> It can be observed that *scientific knowledge advances through rivalry between hypotheses* that explain the same data — for example, the controversy between Millikan and Ehrenhaft helped the development of science (Student #25, Experimental Group B, italics added).
>
> Yes, controversies between scientists can help the development of science — at times the discussions even become bitter. In his controversy with Ehrenhaft, *Millikan always stood firm with his idea of the elementary electrical charge* (Student #16, Experimental Group B, italics added).

It is important to note that many Experimental Group students referred to the Millikan–Ehrenhaft controversy as it formed part of the Study Guide discussed in class within an explicit-reflective framework. This item focused on the relationship between *controversy* and *creativity* within the context of *progress in science*. Now, let us see how some of the students used the Millikan–Ehrenhaft controversy as the background to highlight the following aspects of progress in science:

(a) A scientist that does not use the scientific method may be more creative (Student #28). This sounds reasonable as the scientific method is generally considered to be more rigid and involves a sequence of steps.

(b) Millikan used his creativity to invent the experiment (Student #3).

(c) Millikan–Ehrenhaft controversy provoked the curiosity of the scientific community (Student #7). Indeed, this was very much the case, and although

Millikan was regularly nominated for the Nobel Prize from 1916 onwards, the community recommended that the Prize be withheld until the controversy with Ehrenhaft was resolved. Millikan was finally awarded the Nobel Prize in 1923 (cf. Holton 1988; Niaz 2015).

(d) Millikan used his creativity to discover the oil drop experiment (Student #18).

(e) Scientific knowledge advances through rivalry between hypotheses (Student #25). The item itself makes no mention of the rivalry between hypotheses. However, the importance of rival theories is recognized as an important part of progress in science (Lakatos 1970).

(f) Millikan always stood firm with his idea of the elementary electrical charge (Student #16). Indeed, Millikan's perseverance with his theoretical framework played an important role. For that matter, even Ehrenhaft persevered with his idea of fractional charges as late as the 1940s (Ehrenhaft 1941).

Following are some examples of rhetorical responses provided by Control Group students:

Yes, it can help in the progress of science. For example, if an experiment is performed and the scientist makes mistakes, then he can be corrected by another scientist who did not make mistakes (Student #22, Control Group).

Yes, contact with other scientists can help in the advancement and progress of science (Student #16, Control Group).

No, for example Thomson and Bohr had different theories with respect to some of the experiments performed earlier. Consequently there was confusion as they could not tell which, was the correct theory (Student #32, Control Group).

These responses contrast sharply with those of the Experimental Group students reported earlier. For example, Student #22 refers to a possible controversy if one scientist makes mistakes, then another can correct it. Controversies in the history of science have arisen as two scientists interpret similar data differently and not necessarily due to the mistakes made by one and not the other. In general, such responses are rhetorical based on textbook prescription of one scientist correcting the mistakes of another. Response by Student #32 clearly shows the lack of an historical perspective (with respect to the atomic theories of Thomson, Bohr, and others). In the history of science if one scientist criticizes another theory, it does not necessarily mean that the previous theory was wrong and the new one is correct. In other words, Bohr's theory provided greater explanatory power than that of Rutherford (cf. Lakatos 1970; Niaz 1998, 2009).

Context of Scientific Progress (Students' Responses on Item 5 of Posttest)

The objective of this item was to evaluate students' understanding of the various aspects that are important in the context of scientific progress. Given the experience gained in the experimental treatment and classroom discussions, it is plausible to suggest that Experimental Group students could develop a better understanding of how science progresses.

Table 8 shows that a majority of the Experimental Group students responded conceptually, whereas the Control Group students responded rhetorically. Following are four examples of conceptual responses provided by Experimental Group students:

> The most important aspects of scientific progress are: creativity of the scientists, knowledge of the scientists, controversies between scientists, *scientific knowledge advances through rivalry between hypotheses*, work of one scientist serves as a guide for another (Student #33, Experimental Group B, italics added).
>
> The most important aspects of scientific progress are: there exist a diversity of methods, the experiment of one scientist serves as a base or help for another, there exist controversies or discussions among scientists, results of the experiments need to be published, *work of the scientists is critiqued and corrected and science keeps progressing*, creativity and knowledge helps scientists during the development of an experiment, depending on the nature of the topic scientists base their work on different methods (Student #4, Experimental Group A, italics added).
>
> There exist many controversies among scientists, *one theory is superseded by another*, based on different theoretical frameworks scientists arrive at different conclusions, analyze the same data using different methods (Student #6, Experimental Group B, italics added).
>
> The most important aspects of scientific progress are: criticisms, scientists base their work on the research of other scientists, controversies, discussions, *publication of their research*, use diverse scientific methods (Student #29, Experimental Group A, italics added).

Most of the Experimental Group students emphasized the importance of controversies among scientists and the use of diverse methods. An important feature of these conceptual responses is that at least some of the students expressed the different aspects of scientific progress in their own words and following are some examples:

(a) Scientific knowledge advances through rivalry between hypotheses (Student #33).
(b) Work of the scientists is critiqued and corrected and science keeps progressing (Student #4).
(c) One theory is superseded by another (Student #6).
(d) Publication of their research (Student #29).

The wording of these aspects provides an indication of the degree to which the students changed their epistemological beliefs of NOS. For example, mentioning controversy and then referring to rivalry between hypotheses are important.

Table 8 Comparison of the performance of Control and Experimental Group (A and B) students on Item 5[a] of Posttest

Response	Control ($n = 32$)	Experimental A ($n = 33$)	Experimental B ($n = 38$)
Conceptual	3 (9 %)	28 (85 %)	29 (76 %)
Rhetorical	25 (78 %)	4 (12 %)	4 (11 %)
No response	4 (13 %)	1 (3 %)	5 (13 %)

[a]Item 5: In your opinion, what are the aspects that characterize progress in science?

Associating the idea of 'critiqued' and 'corrected' with progress in science is novel. Another important feature is 'one theory is superseded by another.' The literature in science education generally refers to this aspect as the 'tentative NOS' (cf. Lederman 2004; McComas et al. 1998; Smith and Scharmann 1999).

Following are some examples of rhetorical responses provided by the Control Group students:

> Most important aspects of progress in science are: how to do the experiment, have sufficient material for doing the experiment, which experiment to perform (Student #11, Control Group).
>
> Some of the important aspects are: experiments, observations, technology, benefits, and knowledge about science (Student #20, Control Group).
>
> Some of the important aspects of progress in science are: knowledge of the experiment, what has to be accomplished and an example of what has to be done, have a lot of experience in doing experiments, and have patience in order to be successful (Student #25, Control Group).

Given the information provided in traditional classrooms and textbooks, these responses are not surprising. Actually, it is fairly common even among teachers to suggest the importance of experiments, observations, what has to be done, and patience in order to be successful. Such an understanding of the scientific endeavor should be a cause of concern for teachers, textbook authors, and curriculum developers.

Concept Maps Drawn by Experimental Group Students

Concept maps are an important epistemological tool for facilitating students' conceptual understanding of a science topic (Ausubel et al. 1991; Novak 1990; Novak and Gowin 1988). Experimental Group students were asked to elaborate concept maps during the first week after the instructor had taught the topic in the traditional manner as presented in most textbooks. Students were asked to base the concept map on the following question: How does scientific knowledge develop? Later during the third week after the experimental treatment, students were asked to elaborate another concept map and they based it on the following question: What are the aspects that characterize progress in science?

Concept Maps Drawn by Student #2 (Experimental Group A)

Figure 1 shows the concept map drawn by Student #2 before the experimental treatment. Besides the illustrations (in the original in color) which are quite explicit, this student provided further elaboration of the concept map by including five points (see clockwise). Following is an almost verbatim reproduction of these points (small corrections are included in order to facilitate understanding):

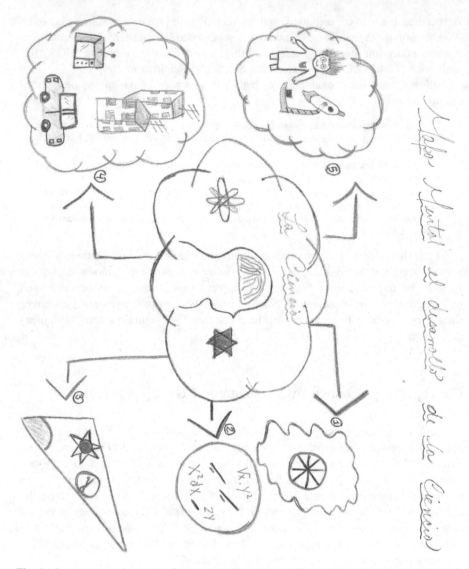

Fig. 1 Concept map drawn by Student #2 (Experimental Group A) before the experimental treatment. Each of the five points (clockwise) in the map is explained further on a separate page (see text for details)

1. From the beginning of the history with the invention of the wheel, the knowledge of human beings started to manifest and this was the beginning of science.
2. Study of science: Archimedes and Pythagoras also contributed toward the development of science based on mathematics, chemistry, and physics.

3. Galileo contributed toward the development of science by studying the planets and the solar system.
4. Objectives: In order to answer everyday questions and improve the quality of life of the people.
5. Through studies in the laboratory and present-day research, science has been able to develop technology for the well-being of humanity.

By any standard, this response to the question, how does scientific knowledge develop, is worthy of merit and recognition. It is thoughtfully constructed and refers to the work of some stalwarts in the history of science (Archimedes, Pythagoras, and Galileo) and the possible objectives pursued by science. Now, let us see how the perspective of this student changes after participating in the experimental treatment (based on the Study Guide and classroom discussions). Figure 2 shows the concept map prepared by this student during the third week after the experimental treatment. Following are some of the salient features of this concept map:

(a) At the center of the concept map is the idea of 'NOS' and from there linkages have been established with the scientific method, science is based on variable methods leading to the advancement of scientific knowledge.
(b) Scientific method is characterized by the following: observations, hypotheses, experiments, results, conclusions, and theory (this, of course, represents the traditional scientific method as presented in most science textbooks and classrooms).
(c) What the scientists use, however, is not the scientific method, but rather the method changes according to the subject under study? Reference is made to the following scientists: Millikan, Faraday, Franklin, Stoney, Thomson, Townsend, and Wilson. This prepares the ground (as suggested in the Study Guide) for the determination of the elementary electrical charge and the ensuing controversy between Millikan and Ehrenhaft.
(d) An explicit mention of how Millikan and Ehrenhaft opposed each other, leading to new knowledge.

Now, let us compare the two concept maps elaborated by Student #2 (Figs. 1 and 2). Most researchers in science education would agree that this student has been able to incorporate in his epistemological beliefs some important aspects of NOS. One of these aspects is that scientists generally do not use the scientific method but rather a variety of methods, which leads to conflicts (Millikan and Ehrenhaft), eventually leading to new knowledge.

Concept Maps Drawn by Student #30 (Experimental Group B)

Figure 3 shows the concept map drawn by Student #30 before the experimental treatment. Following are some salient features of this concept map:

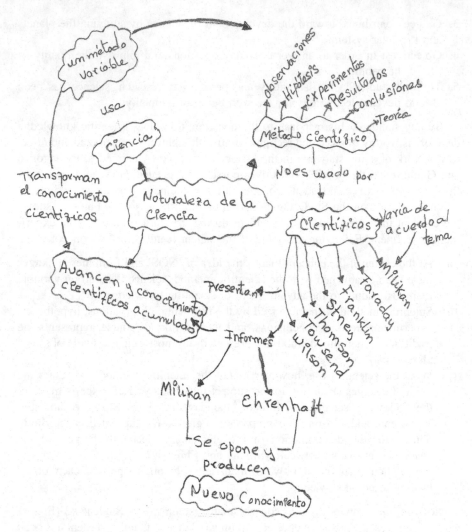

Fig. 2 Concept map drawn by Student #2 (Experimental Group A) after the experimental treatment. In order to facilitate visibility, all the words are retraced with pencil #2. There are some mistakes with respect to names and properties, which were part of the original concept map

1. At the center of the map is 'science.'
2. Linkages are established with (clockwise): provides knowledge, work of the scientists, studies, comfortable products, and benefits.
3. An important aspect of the concept map is that science provides knowledge through the 'thinking mind' (see the human figure).

At this stage, it is interesting to compare the concept map drawn by this student (#30) after the experimental treatment (see Fig. 4). This concept map has the structure of a running dialogue between a student (person on the left) and a teacher

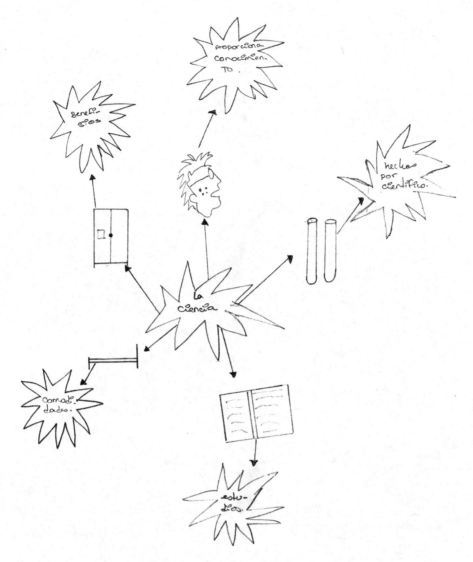

Fig. 3 Concept map drawn by Student #30 (Group B) before the experimental treatment

(person on the right). Actually, it could also be a dialogue between two students. In order to understand better, we present the concept map as three segments involving questions (person on the left) and answers (person on the right):

First segment:

Question Who helps in the development of science?

Answer The scientists.

Question How do the scientists develop science?

Answer Through investigations, experiments, many studies, and then publication.

◀ **Fig. 4** Concept map drawn by Student #30 after the experimental treatment. Strictly speaking, this may not be considered as a concept map. However, it provides insight with respect to student's thinking on the subject and this is what the maps try to accomplish

Second segment:
Question   How is the relationship between the scientists?
Answer   Through controversies, differences, conflicts.
Question   How do the scientists analyze the data?
Answer   Through their own knowledge and creativity.

Third segment:
Question   Is science a reality forever?
Answer   No ... !
Question   Do the scientists manage the reality completely?
Answer   No, science is tentative
Question   Are they (scientists) creative?
Answer   Yes, creativity is very important.

These seven questions subdivided into three segments, in a sense represents the Socratic approach to learning and knowledge. This immediately raises the question, how could a high school student with almost no prior experience in educational research or NOS could elaborate such a dialogue? Interestingly, the concept map drawn by this student before the experimental treatment (Fig. 3) provides no indication of the understanding manifested in Fig. 4. Actually, the structures of the two concept maps (Figs. 3 and 4) are entirely different. This suggests that the classroom treatment provided this student a new perspective that was previously not available. The three questions (Fig. 4) in segment three are thought-provoking indeed and represent some ideas of research methodology (also NOS) in a very lucid manner. The question 'Is science a reality forever?' is impregnated with the obvious answer, and this did not require an extensive response, but instead a laconic 'No ...!' sufficed. Without being overly optimistic, it is plausible to suggest that the experimental treatment played a crucial role in facilitating this students' understanding of the underlying issues. For further elaboration see the last section.

Concept Maps Drawn by Student #9 (Experimental Group B)

Figure 5 shows the concept map drawn by Student #9 before the experimental treatment. The student starts with the central idea: 'the science,' and then moving up and clockwise goes through the following linkages: 'mysteries' → 'scientists work on them' → 'doing experiments' → 'for the advancement of science' → 'leading to our welfare.' This is the traditional understanding of what is science, as construed in most science textbooks and courses. Actually, there is nothing particularly wrong with this perspective of science.

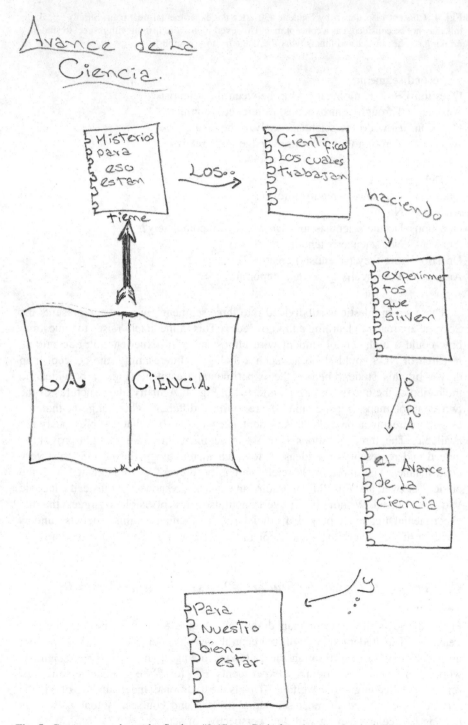

Fig. 5 Concept map drawn by Student #9 (Group B) before the experimental treatment

On the contrary, science education research tries to provide students more informed views of NOS. According to Hodson (2009), 'We should also ask why false or confused NOS knowledge constitutes a major problem for science education. In short, why does it matter what image of science is presented and assimilated? It matters insofar as it influences career choice, and so may have long-term consequences for individuals. It matters if the curriculum image of science is such that it dissuades creative, nonconformist and politically conscious individuals from choosing to pursue science at an advanced level' (p. 142). No wonder, the curriculum image of science conveyed to the students at an early age can have long-term effects.

Now, let us consider the concept map drawn by Student #9 after the experimental treatment (see Fig. 6). The central idea in the concept map has now changed

Fig. 6 Concept map drawn by Student #9 (Group B) after the experimental treatment

from 'the science' (Fig. 5) to 'advancement in science' (Fig. 6). From the central
idea, the student goes up and establishes the following linkages (clockwise): 'a
theory is postulated, however, this theory can change' → 'The scientists are very
creative and use their knowledge very well' → 'Science does not develop alone,
but on the contrary it is accumulative' → 'The controversies among the scientists
makes the study of science more interesting' → 'All the scientists do not use the
scientific method, as the method would depend on the topic of their study and their
creativity.' It is important to note that this student goes through various aspects of
the NOS, without simply reproducing what was discussed in class. The first linkage
(theory can change) refers to the tentative NOS. The second linkage refers to the
creativity of the scientists. The third linkage refers to the accumulative NOS (accumulative here is not used in the philosophical sense, but rather refers to the fact
that scientists do not work alone). The fourth linkage refers to the role played by
controversies. The fifth linkage refers to how the method depends on the topic of
study. In summary, this student has drawn attention to the following aspects of
NOS: tentative, creative, accumulative, controversial, and the scientific method.

Interviews with Experimental Group Students

Seventeen students (selected randomly) from the Experimental Groups (A and B)
were interviewed by the second author. Each interview lasted about 30 min and was
audiotaped and later transcribed. Each interview was based on students' responses
on one or more items of the Pretest and/or Posttest. The researcher showed and read
aloud students' response on a particular item and then asked relevant questions. The
idea behind the interview was to explore students' thinking beyond that expressed
in the written response, and observe any changes. It is important to note that all
interviews were semi-structured and not open-ended, and thus, it was somewhat
difficult to broach issues not included in the item itself or the student's response.
Thus, it was not possible to evaluate long-term retention through the interviews.
A delayed Posttest could have provided some information with respect to students'
retention. However, this was not possible due to class schedule of the students and
course organization followed by the school. Following letters are used to transcribe
the interviews: S = Student and R = Researcher.

Interview with Student #9 (Experimental Group A)

R   What is your opinion with respect to the creativity of the scientist?
S   I consider that in order to develop his/her experiment and knowledge a scientist
     has to be very creative
R   Posttest (Item 5) referred to the important aspects of scientific development and
     you mentioned six, including science is accumulative. Why?

S   It is accumulative, as when the scientists develop their experiments they revise the work of other scientists, and this benefits them mutually

R   On Posttest (Item 1) you stated that there is no unique method for doing science. Why?

S   There exist a diversity of methods and every scientist uses his own method depending on his/her knowledge, experience and creativity

R   On Pretest (Item 1) you stated that Millikan performed his experiment alone without any help. Now you said that science is accumulative. Why did you change your opinion?

S   Well, in the beginning this was not clear, but now I understand

R   In other words, if the teacher only provides the students with the conclusions of the work of a scientist [rhetoric of conclusions], it is not possible to perceive how the development in science really took place

S   Yes, because there are controversies among the scientists and they revise the work of others which leads to confrontations and finally improvement.

Comments:

This student highlights the importance of creativity in the work of the scientists and that they use a diversity of methods. It also clarifies the sense in which the students use the idea of accumulative; for example, Millikan based his work on the contributions of the previous scientists working on the topic. Furthermore, the teacher took the opportunity to clarify that it is not possible to understand development in science by simply presenting rhetoric of conclusions (Schwab 1962, 1974).

Interview with Student #12 (Experimental Group A)

R   In your response to Item 1 (Posttest) you stated that the scientists do not have a unique method of doing science. Why?

S   Yes, because the scientists use their creativity and knowledge in order to choose the method with which they will work. It also depends on the topic on which they are working

R   Considering your response, does it mean that the scientific method does not exist?

S   No, every scientist makes use of his experience and knowledge to determine the method he/she needs. In other words, they use many methods

R   What is more important for a scientist: the data or the theoretical framework?

S   Well, the data

R   So, what do you think happened between Millikan and Ehrenhaft, they had the same data and their conclusions were different?

S   In this case the method made the difference. Millikan did not use the scientific method, whereas Ehrenhaft did. Also they used different theoretical frameworks, which are important for the scientist

R   On Item 4 (Posttest) you stated that the controversy and creativity favor scientific development. Can you amplify further?

S   Yes. Millikan and Ehrenhaft had the same data but they used different theoretical frameworks. They had a confrontation and they started to do study more. Then Millikan used his creativity and experience in order to prove his hypothesis, whereas Ehrenhaft could not do so

R   Why do you think that scientific knowledge is accumulative?

S   Before doing their own experiments, scientists first revise the work of other scientists in order to improve their ideas and develop their experiments

R   Then, do you believe that theories can change?

S   Yes, because the scientists find support in the work of others and the technology also advances. Then the scientists develop their research again, obtaining better results and consequently science advances.

Comments:

This student clarifies that although a unique scientific method does not exist, there are various methods from which the scientists can select the one that suits their experience. Despite being aware of the Millikan–Ehrenhaft controversy, this student first responded that the data were more important than the theoretical framework. Furthermore, this student after clarifying what she/he meant by accumulative goes beyond by explaining that this helps to understand why theories change (tentative NOS).

Interview with Student #20 (Experimental Group A)

R   Do you consider that each of the atomic models adapt to the period in which these were postulated (based on Item 2, Pretest)

S   Yes, because the investigators adapt according to the knowledge that exists when doing their experiments. Thomson postulated a model based on what he knew, and later Rutherford based on his knowledge of alpha and beta particle experiments postulated another model, which contradicted Thomson's model

R   On Item 1 (Pretest) you stated that Millikan developed his oil drop experiment without consulting the work of other scientists. Why?

S   Well, when we studied the experiment for the first time [student is referring to before the experimental treatment] it was only mentioned that Millikan did his experiment and no mention was made of the work of others

R   After the detailed presentation of how Millikan did his experiment, do you still conserve your opinion? [The researcher is referring to the experimental treatment based on the Study Guide]

S   No, now I know that Millikan studied the work of scientists before him and tried to improve. For example, he changed water with oil in order to obtain what he was looking for. Consequently, a scientist looks for support in the previous work of others

R   What do you think is more important for a scientist, his experimental data or theoretical framework? [based on Item 3 of Posttest]

S   The theoretical framework is more important as it helps the scientist to analyze the data and thus provide a correct response

R   Do you consider that creativity and controversy among scientists are important factors in scientific development? [based on Item 4 of Posttest]

S   Yes, because Millikan was being creative when he changed water with oil, helping him to obtain better results and consequently the Nobel Prize

R   Do you believe that the scientific method is the only method that the scientists use? [based on Item 1 of Posttest]

S   No, because every scientist uses the method depending on the experiment he/she performs

R   On Item 3 of the Pretest, you were asked: How does scientific knowledge develop? And you provided no response. Can you provide your opinion now?

S   Yes, some of the important aspects are: methods used by the scientists, creativity of the scientists, controversies among the scientists and their theoretical frameworks. All these lead the scientists to develop their experiments better.

Comments:

This student's responses show clearly how the oil drop experiment is generally presented in textbooks (Caballero and Ramos 2001 in this case) and how this perspective changed after the experimental treatment, based on the Study Guide (see Appendix). Again, the student did not respond to the question (Pretest) that dealt with the development of scientific knowledge and provided a fairly well-elaborated response after the experimental treatment.

Interview with Student #25 (Experimental Group A)

R   Do you think scientists develop their theories alone or do they receive help from other scientists? (based partially on Item 1 of Pretest)

S   Yes, because the work of other scientists helps in going beyond

R   What is your opinion with respect to creativity and controversies among scientists? Do these help in scientific development? (based on Item 4 of Posttest)

S   Yes, it does help. We have the example of Millikan and Ehrenhaft, who had serious disputes and finally Millikan won the 'battle of the electron.' This makes the study more interesting and helped in the advancement of science

R   What do you think is more important for the scientist: experimental data or theoretical framework? (based on Item 3 of Posttest)

S   For me, it is the theoretical framework. Millikan and Ehrenhaft had the same data and it was the theoretical framework and its creative use that helped

R   Who in your opinion was more creative (Millikan or Ehrenhaft)?

S   Well, Millikan, as he did not use the scientific method, whereas Ehrenhaft did

R   Do you believe that there is a unique scientific method for doing science? (based on Item 1 of Posttest)

S   No, not for me, as the method depends on the creativity of each scientist in developing his experiment

R   On Item 5 of Posttest you mentioned that one of the aspects of scientific development is that scientists can change their postulates. Why?

S   Yes. For example, Millikan based on his theoretical framework had postulated an idea [the elementary electrical charge]. However, tomorrow another scientist may come up with another idea that contradicts it [Millikan's]

R   Well, this means that what is reality today may be false tomorrow?

S   Yes, exactly and this is how science advances.

Comments:

First, it is important to note that this student thinks that Millikan was more creative than Ehrenhaft as he did not use the scientific method. Second, this student's response that scientists can change their postulates was clearly unexpected. Although there is some ambiguity in this response, it is clarified later. Student's response to the last question (what is reality today may be false tomorrow?) is categorical: This is how science advances. Actually, the Study Guide does discuss this issue, namely the tentative NOS, in the context of the Thomson, Rutherford, and Bohr models of the atom. However, the issue of changing Millikan's postulates (theoretical framework) was considered beyond the scope of this high school chemistry course and was not discussed.

   Interestingly, scientists have been searching for fractional charges since the 1960s, and that would question Millikan's postulate of the elementary electrical charge. These elementary particles are referred to as 'quarks' and have fractional charges that are multiples of the electron charge. Among others, Martin Perl (Nobel laureate in physics) was until quite recently quite active in isolating quarks (Perl et al. 2004; Perl and Lee 1997). Niaz (2009, Chap. 13) has presented a brief review of the literature related to the search for quarks. At this stage, it is not our intention to suggest that Student #25 was aware of or referring to quarks. However, it is not farfetched to suggest that for this student, if theories are tentative (Thomson, Rutherford, and Bohr models changed in quick succession), then Millikan's postulate (elementary electrical charge) could also change.

Interview with Student #13 (Experimental Group A)

R   Why do you consider scientific knowledge to be accumulative?

S   Yes, because the scientists seek help from the work of other scientists in order to advance their own studies

R Do you consider that there exists a diversity of methods for scientific development? (based on Item 1 of Posttest)

S Because no unique scientific method exists. Besides, the method depends on the nature of the topic under study

R Why do you think there was a difference between the results of Millikan and Ehrenhaft?

S Because Millikan followed his own theoretical framework, whereas Ehrenhaft followed the scientific method and thus included all the drops he had studied without considering that some of them could have had errors. Based on his criteria, Millikan discarded drops that had errors

R Why do you think that creativity, controversy and criticisms among scientists facilitate scientific development? (based on Item 4 of Posttest)

S Yes it helps. If we consider Millikan's experiment it can be observed that besides his knowledge and experience it was his creativity in analyzing the data that helped him to win the Nobel Prize.

Comments:

This student explicitly refers to experimental errors in Millikan and Ehrenhaft's data and stated: 'Ehrenhaft followed the scientific method and thus included all the drops he had studied without considering that some of them could have had errors.' The issue of experimental errors and how it led Millikan to discard data was not explicitly included in the Study Guide or the classroom discussions (these only referred to the fact that Millikan discarded drops that did not provide the expected value of the elementary electrical charge). The role of experimental errors is important, and Holton (1999) refers to them in the following terms: '… Millikan regarded the drops he neglected as unfulfilled, 'aborted' sorts of events on which he thought he need not waste his time to find out what is wrong. (And many things can and do go wrong to prevent a 'reading' to become a 'datum')' (p. 1). So the question we have to deal with is how this student explicitly refers to experimental errors. Indeed, this in a way may represent a creative contribution. More recently, Holton (2014) has gone beyond and clarified further Millikan's discarding of data from oil drops: 'So even if Millikan had included *all* drops and yet come out with the same result, the error bar of Millikan's final result would not have been remarkably small, but large—the very thing Millikan did not like' (p. 1).

   According to Niaz (2005), 'It is plausible to suggest that Ehrenhaft's methodology approximated the traditional scientific method, which did not allow him to discard 'specious drops' [drops with errors]. Millikan, on the other hand, in his publications espoused the scientific method, but in private (see his handwritten notebooks) was fully aware of the dilemma faced and was forced to select data in order to uphold his presuppositions' (p. 699). Interestingly, this student not only refers to experimental errors but also attributes Eherenhaft's selection of 'drops with errors' to his following of the scientific method. In this sense, this is a very interesting response.

Interview with Student #11 (Experimental Group B)

R   Why do you think that the scientific method does not exist and only a diversity of methods characterize scientific development? (based on Item 1 of Posttest)

S   Well, the scientific method exists, but scientists do not follow it strictly. Based on creativity they create their own method in order to develop their research projects

R   You considered creativity and controversy important for scientific development. Why? (based on Item 4 of Posttest)

S   The work of a scientist arouses the curiosity of other scientists and these lead to discussions and controversies that facilitate greater understanding

R   What do you think is more important for a scientist: experimental data or the manner in which it is analyzed? (based on Item 3 of Posttest)

S   Well, the way the data is analyzed is more important, and this is a consequence of his/her creativity and expectations

R   According to what we have studied in this topic of atomic structure, do you think that theories can change? (based on Item 2 of Pretest)

S   Yes, one scientist postulates a theory, and then another scientist comes along and after further investigation the information accumulates and thus the theories keep changing

R   Do you think that the scientists make mistakes or their conclusions adapt the period in which their study is carried out? [This is the essential point the researcher wanted to bring to the student's attention and hence the ground was prepared in the previous questions]

S   Well, really they do not make mistakes. These conclusions depend on the knowledge available in a particular period of time. For example, Thomson did not have the information of alpha particles when he postulated his model. However, Rutherford did have the relevant information and he improved upon Thomson's model.

Comments:

In this interview, the researcher followed a sequence of thought patterns: started with the scientific method → diversity of methods → creativity and controversy → experimental data and how it is analyzed → creativity and expectations (suggested by the student) → can theories change in the context of atomic structure → research continues and theories keep changing. In a sense, this sequence of thought patterns helped the student to understand that scientists do not necessarily make mistakes in their experiments but rather follow the information available to them at a particular period of time and as the information changes the models change. The role played by the expectations of the scientist, as suggested by the student also facilitated understanding of the underlying issues. The actual picture of the dynamics of scientific progress is much more complex, as after Rutherford's alpha particle experiments, Thomson did the same experiments in his own laboratory and disagreed with Rutherford's interpretation—leading to a bitter dispute between the two (for details see Wilson 1983; Niaz 1998, 2009).

It is important to note that this sequence of thought patterns took place within a particular topic of the chemistry curriculum (atomic structure) and this provided the domain-specific context which is essential in order to understand the domain-general aspect of the NOS, namely tentative nature of scientific theories.

Interview with Student #8 (Experimental Group B)

R   In the exam when I asked you if the theories could change you did not respond. Can you provide your opinion now? (The researcher is referring to Item 2 of the Pretest)

S   Yes, theories can change as every scientist has his/her theoretical framework, ideas and knowledge. For example, Thomson postulated an atomic theory, later Rutherford did other studies and did not agree with Thomson and postulated another theory. Similarly, Bohr did the same

R   So, you think that the knowledge is accumulative?

S   Yes, because what one scientist does serves to help the other, either to improve the theory or to change it altogether

R   Why do you think that scientists do not have a unique scientific method for doing science? (based on Item 1 of Posttest)

S   Yes, scientists have various methods and they select one depending on the nature of the topic

R   Do you think that based on his oil drop experiment, Millikan won the Nobel Prize due to his creativity?

S   Yes, this is a very difficult experiment and Millikan had the ability to change water with oil and discard all the drops that he considered unacceptable (with a margin of error) and thus reached a good conclusion

R   We have talked about the theoretical framework of the scientists, the use of diverse methods, controversies and creativity. Do you think that all these aspects are important for scientific development?

S   Yes, all these aspects facilitate scientific development. However, the most important of these is the theoretical framework, as this helps the scientist to develop his ideas.

Comments:

This student did not respond to Item 2 of the Pretest. However, the experience provided by the Study Guide and classroom discussions helped her/him to acquire considerable knowledge with respect to scientific development. Furthermore, the reference to 'unacceptable [drops] (with a margin of error)' is an example of a creative contribution as this was not explicitly discussed in class. For a similar response see interview with Student #13 (Experimental Group A).

Interview with Student #26 (Experimental Group B)

R   Do you think that creativity and controversies help scientists in scientific development? (based on Item 4 of Posttest)

S   Due to the controversies scientists are more careful in presenting their experimental data with respect to a particular topic. In the long run this helps to promote scientific knowledge

R   So, you believe that theories can change? (based on Item 2 of Pretest)

S   Yes, because scientists keep studying the work of others and try to improve upon them. For example, in the case of atomic theory one model was superseded by another

R   Why did Millikan and Ehrenhaft arrive at different conclusions, if they had the same data?

S   Because one of them followed the scientific method strictly and the other did not. Millikan was more successful as he did not use the scientific method and reported data from only those drops that he considered acceptable according to his theoretical framework.

Comments:

This student explicitly relates Millikan's success to having not used the scientific method and thus reported data from only those drops that provided support to his theoretical framework, namely the existence of the elementary electrical charge.

Interviews with Control Group Students

Control Group students were interviewed under the same conditions as the Experimental Group students and were selected randomly (R = Researcher and S = Student).

Interview with Student #8 (Control Group)

R   Why do you think that Millikan did not consider the contributions of other scientists while developing the oil drop experiment?

S   Because, I think if he had shared his ideas, the other scientists could have stolen them

R   Why do you consider the scientific method as important for doing science?

S   Because they have to do experiments while investigating about what is of interest to them

R   Do you think that there exist controversies among scientists?

S   No, there are no controversies among scientists, as this would not allow them to help each other

R   Why do you consider an experiment to be the most important aspect for scientific development?
S   Yes, because it is through experiments that they can think and acquire more knowledge.

Comments:

On the one hand, this student thinks that Millikan did not share his ideas with other scientists and then contradicts by saying that if there were controversies among scientists, then they would not be able to help each other. In a sense, these are spontaneous responses as such ideas are not included in textbooks nor discussed in the classroom. Furthermore, the student thinks that if you do experiments, then you must use the scientific method and experiments themselves constitute the most important aspect of scientific development. This clearly shows that while teaching the oil drop experiment (also other topics and experiments), it would be helpful to include the role played by Millikan's guiding assumptions (theoretical framework) and the controversial nature of experimental data.

Interview with Student #9 (Control Group)

R   Do you consider the scientific method to be the only method used by scientists to do science?
S   Yes, because this is the only method that exists
R   Do you think that theories can change?
S   No, because the scientists guard [defend] their ideas
R   Do you think that controversies among scientists help in the development of science?
S   No, because there are no controversies among them. They propound their theories and each one respects the ideas of others
R   Is creativity important for scientific development?
S   Yes, because this helps in the development of science with more precision
R   What aspects are most important for the development of science?
S   The materials used by the scientists and the data they handle.

Comments:

For this student, the scientific method is the only method that exists, theories do not change, there are no controversies among scientists, and experimental data are the most important aspect of scientific development. For anyone familiar with school science textbooks and curriculum (in different parts of the world), these ideas are quite understandable.

Interview with Student #19 (Control Group)

R   Do you consider that scientists only use the scientific method while doing their research?
S   Because this is what they do: collect experimental data, analyze it and then report the results
R   Why do you consider the experimental data to be more important than the theoretical framework of the scientists?
S   Because the data are provided by the experiments
R   When we studied atomic theory did you observe any controversy among the scientists? If controversies exist, do you think they help in the development of science?
S   No, there was no controversy. If it did, that would not be helpful, as they will start discussing among themselves instead of doing what they had to do
R   Why do you think that Millikan did not consider the contributions of other scientists while doing the oil drop experiment?
S   Because, he was concerned about proving his own ideas
R   Which aspects of scientific development do you consider to be the most important?
S   Through their experiments scientists look for technological advances that can cure diseases and thus provide the people with a better quality of life.

Comments:

Responses of this student are quite representative of most high school students in many parts of the world (cf. Dogan and Abd-El-Khalick 2008 for Turkish high school students). Following are some of the salient features: (a) The work of a scientist based on the scientific method consists of collecting data, analyzing it and then reporting it. This simplifies the scientific endeavor to a simple stepwise procedure and should be a cause of concern if we want our future scientists to have a deeper understanding of the complexities involved; (b) it seems as the data are provided by experiments, the latter are infallible. Most historical episodes demonstrate how the design and the subsequent interpretation of experimental data inevitably lead to alternatives and controversies; (c) a simple presentation of the atomic models (Thomson, Rutherford, Bohr, others) does not help students to understand that if the models changed, there must have been some reason for it and consequently the possibility of controversies. Furthermore, the student thinks that the controversies may distract the scientists from doing what they had to do; (d) it is interesting that the student considers the technological advances and a better quality of life as important aspects of scientific development. Of course, this ignores the scientific endeavor itself and deals with the possible benefits of science.

References

Akerson, V. L., Abd-El-Khalick, F., & Lederman, N. G. (2000). Influence of a reflective, explicit activity-based approach on elementary teachers' conceptions of nature of science. *Journal of Research in Science Teaching, 37*, 295–317.

Alexander, P. A. (1992). Domain knowledge: Evolving issues and emerging concerns. *Educational Psychologist, 27*, 33–51.

Ausubel, D., Novak, J., & Hanesian, H. (1991). *Psicología Educativa: Un punto de vista cognoscitivo*. México, D.F.: Trillas.

Caballero, A., & Ramos, F. (2001). *Química: Teoría, problemario, auto evaluación* (7th ed.). Caracas: Distribuidora Escolar.

Dogan, N., & Abd-El-Khalick, F. (2008). Turkish grade 10 students' and science teachers' conceptions of nature of science: A national study. *Journal of Research in Science Teaching, 45*(10), 1083–1112.

Ehrenhaft, F. (1941). The microcoulomb experiment. *Philosophy of Science, 8*, 403–457.

Hodson, D. (2009). *Teaching and learning about science: Language, theories, methods, history, traditions and values*. Rotterdam: Sense Publishers.

Holton, G. (1969). Einstein and the 'crucial' experiment. *American Journal of Physics, 37*, 968–982.

Holton, G. (1988). On the hesitant rise of quantum physics research in the United States. In S. Goldberg & R. H. Stuewer (Eds.), *The Michelson era in American science, 1870–1930* (pp. 177–205). New York: American Institute of Physics.

Holton, G. (1999). Personal communication to the first author, April 29.

Holton, G. (2014). Personal communication to the first author, August 3.

Lakatos, I. (1970). Falsification and the methodology of scientific research programmes. In I. Lakatos & A. Musgrave (Eds.), *Criticism and the growth of knowledge* (pp. 91–195). Cambridge, UK: Cambridge University Press.

Lederman, N. G. (2004). Syntax of nature of science within inquiry and science instruction. In L. B. Flick & N. G. Lederman (Eds.), *Scientific inquiry and nature of science* (pp. 301–317). Dordrecht, The Netherlands: Springer.

Lederman, N. G., & O'Malley, M. (1990). Students' perceptions of tentativeness in science: Development, use, and sources of change. *Science Education, 74*, 225–239.

López, J. B. (2006). El enlace covalente y el experimento de Millikan, desde el punto de vista de la historia y filosofía de la ciencia, en libros de texto del primer año de ciencias del ciclo diversificado. Master of Science thesis (Chemistry education). Universidad de Oriente, Cumaná, Venezuela.

Machamer, P., Pera, M., & Baltas, A. (2000). Scientific controversies: An introduction. In P. Machamer, M. Pera, & A. Baltas (Eds.), *Scientific controversies: Philosophical and historical perspectives* (pp. 3–17). New York: Oxford University Press.

McComas, W. F. (2008). Seeking historical examples to illustrate key aspects of the nature of science. *Science & Education, 17*(2), 249–263.

McComas, W. F., Almazroa, H., & Clough, M. P. (1998). The role and character of the nature of science in science education. *Science & Education, 7*, 511–532.

Niaz, M. (1998). From cathode rays to alpha particles to quantum of action: A rational reconstruction of structure of the atom and its implications for chemistry textbooks. *Science Education, 82*, 527–552.

Niaz, M. (2000). The oil drop experiment: A rational reconstruction of the Millikan-Ehrenhaft controversy and its implications for chemistry textbooks. *Journal of Research in Science Teaching, 37*(5), 480–508.

Niaz, M. (2005). An appraisal of the controversial nature of the oil drop experiment: Is closure possible? *British Journal for the Philosophy of Science, 56*, 681–702.

Niaz, M. (2009). *Critical appraisal of physical science as a human enterprise: Dynamics of scientific progress*. Dordrecht, The Netherlands: Springer.

Niaz, M. (2015). Myth 19: That the Millikan oil-drop experiment was simple and straightforward. In R. L. Numbers & K. Kampourakis (Eds.), *Newton's apple and other myths about science* (pp. 157–163). Cambridge, MA: Harvard University Press.

Niaz, M., & Coştu, B. (2013). Analysis of Turkish general chemistry textbooks based on a history and philosophy of science perspective. In M. S. Khine (Ed.), *Critical analysis of science textbooks: Evaluating instructional effectiveness* (pp. 199–218). Dordrecht, The Netherlands: Springer.

Niaz, M., Herron, J. D., & Phelps, A. J. (1991). The effect of context on the translation of sentences into algebraic equations. *Journal of Chemical Education, 68*, 306–309.

Novak, J. D. (1990). Concept mapping: A useful tool for science education. *Journal of Research in Science Teaching, 27*(10), 937–949.

Novak, J., & Gowin, B. (1988). *Aprendiendo a aprender*. Madrid: Martínez Roca.

Osborne, J., Collins, S., Ratcliffe, M., Millar, R., & Duschl, R. (2003). What 'ideas-about-science' should be taught in school science? A Delphi study of the expert community. *Journal of Research in Science Teaching, 40*, 692–720.

Perl, M. (2007). *A contrarian view of how to develop creativity in science and engineering*. Paper presented at The Eighth Olympiad of the Mind, The National Academies, Washington, DC., November (SLAC-PUB-12850).

Perl, M., & Lee, E. R. (1997). The search for elementary particles with fractional electric charge and the philosophy of speculative experiments. *American Journal of Physics, 65*, 698–706.

Perl, M., Lee, E. R., & Loomba, D. (2004). A brief review of the search for isolatable fractional charge elementary particles. *Modern Physics Letters A, 19*, 2595–2610.

Schwab, J. J. (1962). *The teaching of science as enquiry*. Cambridge, MA: Harvard University Press.

Schwab, J. J. (1974). The concept of the structure of a discipline. In E. W. Eisner & E. Vallance (Eds.), *Conflicting conceptions of curriculum* (pp. 162–175). Berkeley, CA: McCutchan Publishing Corp.

Smith, M. U., & Scharmann, L. C. (1999). Defining versus describing the nature of science: A pragmatic analysis for classroom teachers and science educators. *Science Education, 83*(4), 493–509.

Wilson, D. (1983). *Rutherford: Simple genius*. Cambridge, MA: MIT Press.

Conclusions and Educational Implications

The study reported here is based on a reflective, explicit, and activity-based approach to introducing nature of science (NOS) in the classroom that facilitates an understanding of scientific progress (Akerson and Volrich 2006). Results obtained show that the difference in the performance (conceptual responses) of the Control and Experimental Group (A and B) students on the three items of the Pretest is statistically not significant. This means that both groups had a very similar preparation on topics that are relevant for this study. However, on the five items of the Posttest, the difference in the performance of Control and Experimental Groups on conceptual responses is statistically significant (chi-square, $p < 0.01$).

Experimental Group students in this study participated in the following activities and thus had considerable opportunity to familiarize themselves with the topic of atomic structure: *first week*, instruction of Thomson, Rutherford, and Bohr models of the atom and the oil drop experiment based on a traditional format. Next, they discussed this content with the instructor and then prepared concept maps; *second week*, application of the 3-item Pretest and distribution of the Study Guide with preliminary instructions in order to prepare for the following week; *third week*, discussion among the students (with the instructor as moderator) with respect to the material in the Study Guide, elaboration of concept maps; *fourth week*, application of the 5-item Posttest; and *after the fourth week*, semi-structured interviews with students that were selected randomly.

On Item 1 of the Posttest, most of the Control Group students emphasized the scientific method and had a rhetorical response. On the other hand, conceptual responses also acknowledged that scientists use some form of method, which is, however, moderated (accompanied) by the previous experience and creativity of the scientist. In a sense, rhetorical responses are not necessarily incorrect but rather obscure what the scientist actually did. Item 2 of the Posttest dealt with the astrophysicists and was meant to evaluate Experimental Group students' ability to transfer their experience from the oil drop experiment to the present context. Results obtained show that the performance of the Experimental Groups (A and B)

© The Author(s) 2016
M. Niaz and M. Rivas, *Students' Understanding of Research Methodology
in the Context of Dynamics of Scientific Progress*, SpringerBriefs in Education,
DOI 10.1007/978-3-319-32040-3_4

was about as high as on Item 1 and thus these students could transfer knowledge from one context to another (i.e., from Item 1 to Item 2). Interestingly, a small group of Control Group students were also able to transfer knowledge, and this can be attributed to the context effect, namely the context of the problem helps the students to make sense of the problem situation. Item 3 of the Posttest asked students about the importance of experimental data for scientists. The item itself makes no mention of theoretical frameworks, contradictions, conflicts, and controversies. However, many Experimental Group students referred to the following aspects in their responses: Data can lead to contradictions and conflicts, data constitute evidence for hypotheses, creativity, and knowledge of the scientist helps in understanding data, a theoretical framework can help a scientist to foresee where the work is heading, scientists can have the same experimental data, and still their theoretical frameworks can be entirely different. This clearly shows that the Experimental Group students are not simply reproducing what was discussed in class (based on the Study Guide), but rather adding new elements in their efforts to understand the difference between experimental data and theoretical frameworks. Item 4 focused on the relationship between controversy and creativity within the context of progress in science. Responses of Experimental Group students demonstrated greater understanding by referring to the following aspects: Lack of creativity may be due to the use of the scientific method (see response of Student #28 on Item 4), Millikan's use of creativity to invent the experiment (see response of Student #3 on Item 4), Millikan–Ehrenhaft controversy provoked the curiosity of the scientific community (see response of Student #7 on Item 4), Millikan used his creativity to discover the experiment (see response of Student #18 on Item 4), scientific knowledge advances through rivalry between hypotheses (see response of Student #25 on Item 4), and Millikan persevered with his theoretical framework (see response of Student #16 on Item 4). This last aspect is important in understanding scientific progress as history of science shows that scientists generally do not abandon their theoretical frameworks when faced with the first signs of anomalous data (cf. Lakatos 1970). Item 5 of Posttest asked the students to characterize progress in science. Many students expressed the different aspects of scientific progress in their own words and context, and following are some examples: scientific knowledge advances through rivalry between hypotheses, work of the scientists is critiqued and corrected and science keeps progressing, one theory is superseded by another, and publication of their research. For most high school students, the idea of a scientist's work being critiqued, corrected, and then published is quite novel.

Multiple Data Sources

An important feature of this study is the use of multiple data sources. Findings in this study are supported by the following multiple data sources: (a) written responses from Control and Experimental Group students on eight items that formed part of Pretest (3 items) and Posttest (5 items). All the items were open-ended, and thus, the students were not constrained by the test format. (b) Concept maps constructed by the students before the Pretest and again after the experimental treatment (before the Posttest) were particularly helpful in facilitating understanding. (c) Semi-structured interviews with eleven Control Group and seventeen Experimental Group students provided greater insight into students' written responses on the Pretest and Posttest and also gave the students the possibility to include and elaborate new information. Comparison of the responses on test items, concept maps, and the interviews provided considerable depth to the findings of this study. Working with multiple data sources approximates to *triangulation of data sources* and has been endorsed by Johnson and Onwuegbuzie (2004): 'Researchers should collect multiple data using different strategies, approaches and methods in such a way that the resulting mixture or combination is likely to result in complementary strengths' (p. 18). Similarly, Guba and Lincoln (1989) have endorsed the triangulation of multiple data sources: 'Triangulation should be thought of as referring to cross-checking specific data items of a factual nature (number of target persons served, number of children enrolled in a school-lunch program …' (p. 24). In this study, the data items of a factual nature would be the following: (a) number of students who responded conceptually or rhetorically on the eight items of the Pretest and Posttest; (b) elaboration of linkages in the concept maps; and (c) addition of new information during the interviews.

How Concept Maps Can Facilitate Socratic Thinking

Most students were quite enthusiastic about the elaboration of concept maps and dedicated considerable time and effort in expressing what they considered to be the underlying issues. The concept map drawn by Student #30 (Fig. 4) after the experimental treatment is a good example of the student's interest, engagement with the topic, and creativity. It could even be considered as representing the Socratic approach to learning and education. Nola (1997) a philosopher of science, in the context of recent trends in constructivism has referred to the Socratic approach in the following terms:

> In Socrates' view, students do not acquire knowledge through picking up bits of (true) information didactically conveyed to them. Even being led through a question-answer session does not provide, by itself, knowledge; at best the process can only lead pupils to the correct belief. Only when they can go through the steps of reasoning by themselves and thereby make fully explicit to themselves the reasons for the correct answer will they have knowledge (p. 59).

We wonder, how would Nola characterize this concept map (Fig. 4)? Interestingly, this student is perhaps playing the role of Socrates and the student at the same time. At one stage of the concept map (third segment), the student asked a very thought-provoking question: 'Is science a reality forever?' Indeed, this is one of the most difficult ideas related to nature of science and the dynamics of scientific progress. It is interesting to consider how this student arrived at this question? Was it the experimental treatment, the classroom discussions or his/her own curiosity and the ability to go beyond what was discussed in class? Niaz et al. (2003) have argued that progress in science and educational theory (constructivism) is characterized by continual critical appraisals. Physicist–philosopher of science, Holton (1986) has expressed this idea in cogent terms:

> ... the scientists chief duty ... [is] ... not the production of a flawlessly carved block, one more in the construction of the *final Temple of science*. Rather, it is more like participating in a building project that has no central planning authority, where no proposal is guaranteed to last very long before being modified or overtaken, and where one's best contribution may be one that furnishes a plausible base and useful material for the next stage of development (p. 173, italics added).

Indeed, these ideas characterize the research methodology used by scientists in the context of the dynamics of scientific progress. This, of course, leads to the crucial question, how often do we convey this message and facilitate such thinking in our educational practice? On the contrary, most science educators and textbook authors prepare students as if they were going to enter the '*final Temple of science*.'

Concept maps drawn by most Experimental Group students before the experimental treatment were reproductions of how science is depicted in most high school textbooks and courses (also Control Group students drew very similar maps). After the experimental treatment, most students changed their perspective and drew concept maps in which they emphasized the creative, accumulative, controversial nature of science, and the scientific method. Of course, students' use of accumulative in this context is somewhat problematic as a simple accumulation of data means nothing, unless accompanied by heuristic principles as suggested by Schwab (1962), also see Niaz (2012, p. 211).

Changing Nature of Students' Understanding of Progress in Science Based on Interviews with Experimental Group Students

Interviews with Experimental Group students provided a good opportunity to observe how students' thinking changed after the experimental treatment. Following are some of the salient aspects referred to by the students:

(a) Learning just the conclusions related to a topic (rhetoric of conclusions) does not help to understand how science really progresses.
(b) Accumulative nature of science based on the contributions of various scientists facilitates an understanding of the tentative nature of science.
(c) In the oil drop experiment, Millikan used his creativity by using oil instead of water.

(d) Millikan was more creative as he did not use the scientific method. Classroom discussions only referred to the fact that the scientific method is not very helpful in doing science. The relationship between creativity and lack of scientific method was introduced based on the interpretation of some of the students.

(e) Due to the possibility of controversies, scientists are more careful in presenting their experimental data (even Millikan followed this advice, see Holton 1978).

(f) If science is tentative, then even Millikan's theoretical framework based on the elementary electrical charge could also change. This aspect was not discussed in class, and its inclusion adds a new dimension in students' understanding of scientific progress.

(g) Ehrenhaft ignored the fact that some of the drops may have had experimental errors. Classroom discussions had emphasized that as compared to Millikan, Ehrenhaft had included all the drops and hence the wide range of charges observed by him. The reference to experimental errors is important as this is precisely why Millikan's interpretation of data was eventually accepted by the scientific community.

(h) Sequence of thought patterns and scientists' expectations helped the students' to understand that scientists do not necessarily make mistakes while doing the experiments. Models change as the information available at a particular period of time changes.

At this stage, it would be interesting to compare the interviews between the Control and Experimental Group students. In contrast to Experimental Group students, Control Group students expressed the belief that as follows: (i) Experiments constitute the most important aspect of scientific development and that theories do not change; (ii) a step-wise scientific method helps the scientist to collect data, analyze it, and then report it; (iii) experimental data are the final arbiter in scientific development, and hence, there are no controversies in science; and (iv) even if atomic models change, it does not mean that science is tentative.

History of Science Is 'Inside' Science

Many science teachers complain that the curriculum is already very lengthy which makes its coverage very difficult within the time period allocated and hence the inclusion of history and philosophy of science (HPS) in the classroom is not a feasible project. In contrast, Bevilacqua and Bordoni (1998) have stated that: 'We are not interested in adding the history of physics to teaching physics, as an optional subject: the history of physics is 'inside' physics' (p. 451). Matthews (1998) has argued that philosophy is not far below the surface in any science classroom, as most textbooks and classroom discussions deal among others, with concepts, such as law, theory, model, explanation, cause, hypothesis, confirmation, observation, evidence, and idealization (p. 168). Similarly, Niaz and Rodríguez (2001) based on a historical framework have shown that HPS is already 'inside' chemistry and we do not need separate courses for its introduction. Paraskevopoulou and Koliopoulos (2011) considered as an advantage of their study: '... the ability to apply this teaching strategy [for teaching Millikan-Ehrenhaft controversy] even in the context of traditional forms of education in the natural sciences without great changes to the curriculum, something that can encourage teachers to choose to teach NOS aspects in a more systematic way' (p. 957). Similarly, the present study also found that the controversy between Millikan and Ehrenhaft and the related NOS aspects can easily be included in the traditional chemistry curriculum without needing extra class time.

Finally, it is concluded that a teaching strategy based on a reflective, explicit, and activity-based approach in the context of the oil drop experiment can facilitate high school students' understanding of how scientists elaborate theoretical frameworks, design experiments, report data, and elaborate alternative interpretations of data that lead to controversies and finally, with the collaboration of the scientific community, a consensus is reached.

References

Akerson, V. L., & Volrich, M. L. (2006). Teaching nature of science explicitly in a first-grade internship setting. *Journal of Research in Science Teaching, 43*, 377–394.

Bevilacqua, F., & Bordoni, S. (1998). New contents for new media: Pavia project physics. *Science & Education, 7*, 451–469.

Guba, E. G., & Lincoln, Y. S. (1989). *Fourth generation evaluation*. Newbury Park, CA: Sage.

Holton, G. (1978). Subelectrons, presuppositions, and the Millikan-Ehrenhaft dispute. *Historical Studies in the Physical Sciences, 9*, 161–224.

Holton, G. (1986). *The advancement of science and its burdens*. Cambridge, UK: Cambridge University Press.

Johnson, R. B., & Onwuegbuzie, A. J. (2004). Mixed methods research: A research paradigm whose time has come. *Educational Researcher, 33*, 14–26.

Lakatos, I. (1970). Falsification and the methodology of scientific research programmes. In I. Lakatos & A. Musgrave (Eds.), *Criticism and the growth of knowledge* (pp. 91–195). Cambridge, UK: Cambridge University Press.

Matthews, M. R. (1998). In defense of modest goals when teaching about the nature of science. *Journal of Research in Science Teaching, 35*, 161–174.

Niaz, M. (2012). *From 'science in the making' to understanding the nature of science: An overview for science educators*. New York: Routledge.

Niaz, M., Abd-El-Khalick, F., Benarroch, A., Cardellini, L., Laburú, C.E., Marín, N., et al. (2003). Constructivism: Defense or a continual critical appraisal—A response to Gil-Pérez, et al. *Science & Education, 12*, 787–797.

Niaz, M., & Rodríguez, M. A. (2001). Do we have to introduce history and philosophy of science or is it already 'inside' chemistry? *Chemistry Education: Research and Practice in Europe, 2*, 159–164.

Nola, R. (1997). Constructivism in science and science education: A philosophical critique. *Science & Education, 6*, 55–83.

Paraskekevopoulou, E., & Koliopoulos, D. (2011). Teaching the nature of science through the Millikan-Ehrenhaft dispute. *Science & Education, 20*(10), 943–960.

Schwab, J. J. (1962). *The Teaching of Science as Enquiry*. Cambridge, MA: Harvard University Press.

Appendix

Study Guide Based on the Millikan–Ehrenhaft Controversy

Robert Millikan obtained his doctorate from the University of Columbia in 1895 at the age of 27. In 1896, he accepted an invitation to join the physics department at the University of Chicago and became involved in teaching advanced courses on electron and kinetic theory. Millikan drew inspiration from the early work of Franklin, Faraday, Stoney, Thomson, Townsend, and Wilson to develop the idea of the existence of an elementary electrical charge, and this later became the theoretical framework for the oil drop experiment. Millikan's work started with a critical review of the work of Townsend and Thomson. Next, Wilson determined the elementary electrical charge by studying clouds of charged water droplets moving in electrical and gravitational fields. Millikan improved upon Wilson's method by using an electric field strong enough to disperse the cloud of water droplets and leaving a small number of water droplets that could be observed with much ease. A major source of error at this stage was the gradual evaporation of the water droplets as it was difficult to hold the droplets under observation for more than a minute. To avoid this and other problems, Millikan substituted water with oil which eventually led him to measure the charge of the electron.

The oil drop experiment is difficult to perform in the laboratory. Many years later Holton compared Millikan's published results with his laboratory notebooks. It was found that due to the complexity of the experimental conditions, Millikan discarded data from oil drops that did not have velocities within a certain range. Of the 140 drops in the notebooks, in his publication, Millikan reported data from only 58 drops that he considered to be in the correct range.

Felix Ehrenhaft studied at the University of Vienna and the Institute of Technology at Vienna. He was accepted as privatdocent at the University of Vienna in 1905 and taught statistical mechanics. Ehrenhaft was about 10 years younger than Millikan and by 1910 was a fairly well-established figure in the European scientific community. Ehrenhaft's determination of electrical charges was based on the preparation of colloids and the ultramicroscopic Brownian movements of observations of individual fragments of metals such as those from the vapor of a

© The Author(s) 2016
M. Niaz and M. Rivas, *Students' Understanding of Research Methodology in the Context of Dynamics of Scientific Progress*, SpringerBriefs in Education, DOI 10.1007/978-3-319-32040-3

silver arc. By measuring the motions of colloidal particles with and without a horizontal electrical field and applying Stoke's law, he measured the charges on the particles. In contrast to Millikan, he did not use a vertical electrical field. A major shortcoming of this method was that observations were based on two different drops, one for observing the particles without the electric field and the other with the electric field. In 1910, Ehrenhaft conducted new studies in which he used a vertical field strong enough to make particles rise against gravitation (similar to Millikan's method).

The controversy between Millikan and Ehrenhaft started in February 1910, with Millikan's first major publication in the *Philosophical Magazine*, in which he criticized Ehrenhaft's method. This came to be known as the 'battle over the electron,' and the dispute became bitter for the next fifteen years, which led the scientific community to look closely at the experimental data of both Millikan and Ehrenhaft. According to Millikan's theoretical framework, there existed a fundamental elementary electrical charge and the charges on the oil drops were whole number multiples of this fundamental charge. On the other hand, according to Ehrenhaft's theoretical framework, charges on the drops varied considerably, and hence, a fundamental electric charge did not exist.

Both Millikan and Ehrenhaft had very similar experimental data, which they analyzed by methods based on their respective theoretical frameworks. Millikan did not follow the scientific method as he discarded data from drops that did not have velocities within a certain range. Furthermore, despite obtaining anomalous data, he continued to work with his theoretical framework. On the other hand, Ehrenhaft strictly followed the steps of the scientific method and, despite obtaining anomalous data, included all the drops, thus maintained his theoretical framework and ignored the experimental variables that affected the properties of the drops. Handling of the experimental data by the two scientists shows that a unique scientific method does not exist and the same data can be interpreted in more than one way. This shows that experimental data is important for the scientists, but their theoretical frameworks are even more important. After many years of the controversy, finally the scientific community recognized Millikan's creativity in handling the data and he was awarded the Nobel Prize in 1923. This shows the importance of creativity in scientific development and how it enables scientists to go beyond existing knowledge.

Scientists are not solitary geniuses but rather depend on the work of other scientists. In the case of the oil drop experiment, Millikan started his work by a critical evaluation of the previous work of Townsend, Thomson, and Wilson based on charged clouds of water droplets. With this background, we can also understand better the atomic models postulated by Thomson, Rutherford, and Bohr. Rutherford critiqued and went beyond Thomson. Similarly, Bohr questioned and explained the difficulties involved in Rutherford's model of the atom. All scientific knowledge involves contradictions and conflicts that lead to the postulation of new theories. These examples also show that scientific knowledge advances by the rivalry between competing hypotheses and theoretical frameworks.

Printed in the United States
By Bookmasters